Psychosocial Explorations of Film and Television Viewing

Studies in the Psychosocial

Edited by:

Peter Redman, The Open University, UK

Stephen Frosh, Centre for Psychosocial Studies, Birkbeck College, University of London, UK

Wendy Hollway, The Open University, UK

Titles include:

Stephen Frosh
HAUNTINGS: PSYCHOANALYSIS AND GHOSTLY TRANSMISSIONS

Uri Hadar
PSYCHOANALYSIS AND SOCIAL INVOLVEMENT
Interpretation and Action

Margarita Palacios
RADICAL SOCIALITY
Studies on Violence, Disobedience and the Vicissitudes of Belonging

Derek Hook
(POST)APARTHEID CONDITIONS

Gath Stevens, Norman Duncan and Derek Hook (*editors*)
RACE, MEMORY AND THE APARTHEID ARCHIVE
Towards a Transformative Psychosocial Praxis

Irene Bruna Seu
PASSIVITY GENERATION
Human Rights and Everyday Morality

Lynn Chancer and John Andrews (*editors*)
THE UNHAPPY DIVORCE OF SOCIOLOGY AND PSYCHOANALYSIS

Kate Kenny and Marianna Fotaki (*editors*)
THE PSYCHOSOCIAL AND ORGANIZATION STUDIES
Affect at Work

James S. Ormrod
FANTASY AND SOCIAL MOVEMENTS

Jo Whitehouse-Hart
PSYCHOSOCIAL EXPLORATIONS OF FILM AND TELEVISION VIEWING
Ordinary Audience

Studies in the Psychosocial Series
Series Standing Order ISBN 978–0–230–30858–9 (hardback)
978–0–230–30859–6 (paperback)
(outside North America only)

You can receive future titles in this series as they are published by placing a standing order. Please contact your bookseller or, in case of difficulty, write to us at the address below with your name and address, the title of the series and one of the ISBNs quoted above.

Customer Services Department, Macmillan Distribution Ltd, Houndmills, Basingstoke, Hampshire RG21 6XS, England

---

Psychosocial Explorations of Film and Television Viewing

Ordinary Audience

Jo Whitehouse-Hart

palgrave
macmillan

First published 2014 by
PALGRAVE MACMILLAN

Palgrave Macmillan in the UK is an imprint of Macmillan Publishers Limited, registered in England, company number 785998, of Houndmills, Basingstoke, Hampshire RG21 6XS.

Palgrave Macmillan in the US is a division of St Martin's Press LLC, 175 Fifth Avenue, New York, NY10010.

Palgrave Macmillan is the global academic imprint of the above companies and has companies and representatives throughout the world.

Palgrave® and Macmillan® are registered trademarks in the United States, the United Kingdom, Europe and other countries

ISBN: 978–0–230–36283–3

This book is printed on paper suitable for recycling and made from fully managed and sustained forest sources. Logging, pulping and manufacturing processes are expected to conform to the environmental regulations of the country of origin.

A catalogue record for this book is available from the British Library.

Library of Congress Cataloging-in-Publication Data

Whitehouse-Hart, Jo.
Psychosocial explorations of film and television viewing : ordinary audience / Jo Whitehouse-Hart.
pages cm. – (Studies in the psychosocial)
ISBN 978–0–230–36283–3 (hardback)
1. Mass media – Audiences – Psychology. 2. Motion picture audiences-Psychology. 3. Television viewers – Psychology. 4. Motion pictures – Psychological aspects. 5. Television – Psychological aspects. I. Title.

P96.A83W525 2014
302.23'43—dc23 2014024825

For Andy, Isaac and Albert

Contents

Acknowledgements

There are a number of people who must be thanked for their help and support as without them this project would not have been completed. To Andy Hart, without your endless stream of cups of tea and snacks and willingness to pitch in with absolutely anything that needed doing, I would never have been able to finish this. To my wonderful sons Isaac and Albert, thanks for your patience and support, hugs, neck rubs, cups of redbush tea, making me laugh and just generally being super boys. Sorry you have not had a dining table for months.

There are a number of friends and colleagues who have read drafts and given comments at various stages. I am particularly grateful to Richard Johnson whose comments and support have been really helpful. Thanks to dear friends for help and support above and beyond the call of duty, in particular Diane Taylor, Stuart Price, Tim O'Sullivan, Stuart Hanson (whose humour lifts me every day), Helen Wood, Margaret Montgomerie and Vic Blake. I also need to thank two great PhD students for their practical help; Claire Sedgewick and Jilly Kay. Thanks also to Libby Forrest and Nicola Jones at Palgrave for your efficiency and patience and series editors Stephen Frosh and Wendy Hollway for your helpful comments.

Lesley Fielding, Elizabeth Gooderham, Vivian Richardson, Jo Bannon, Harry and Phyllis Tetley (you send the most encouraging text messages), Marianne and Rob Stone, Suzanne and Duncan Groocot, Andy Childerhouse, your support has been invaluable and I couldn't have done it without you.

A very special thank you must go to the wonderful viewers who were willing to give up their precious time to talk and share their life stories, thoughts and feelings with me. I am more than grateful; I am privileged to have been given such a special gift.

Thanks to Karnac books for permission to reproduce a version of Chapter 3 which was published in 2014 as "Spending Too Much Time Watching TV?" in Yates, C., Bainbridge C. and Ward, I. (eds) *Psychoanalysis and Television, Psychoanalysis and Popular Culture* series, London, Karnac.

A version of Chapter 6 was previously published as: "Surrendering to the Dream: An Account of the Unconscious Dynamics of a Research Relationship" *International Journal of Research Practice* special edition Psychodynamic research, Vol. 8, No. 2.

Introduction: Puzzling Viewing

A viewing experience

A young mother sits with a newborn sleeping child on her lap. Channel hopping, she falls on an afternoon showing of a favourite Hollywood musical she has seen so many times she has lost count. In fact she knows bits of the script word for word. The film is *Gypsy* (1962, Dir. Jerome Robbins) and tells the story of *Gypsy Rose Lee*, considered to be the most famous stripper to emerge from Burlesque theatre, and the relationship with her pushy stage mother, *Rose*. This is a film she has only ever seen on television. Her viewing is surrounded, therefore, by the familiar things that are associated with home and where the emotional traces of family relationships imbue the setting affectively. Memories of watching the film as a child with her own mother emerge. She feels uplifted at songs about "getting out" and "everything coming up roses" and remembers wanting to "get out" of her own working class background. Instantly, this viewer becomes re-immersed in the fascinating relationship between aspirational *Rose* and her daughter *Louise/Gypsy Rose Lee*. *Gypsy* "makes it", but not in the way *Rose* had wanted, in legitimate theatre, but in the shameful and seedy arena of burlesque, as a stripper. Ironically, *Gypsy's* material success allows her to dress in beautiful clothes, become cultured, take French lessons and present herself as a "fine lady". The relationship between mother and daughter disintegrates as *Rose*, full of envy for her daughter, and mourning the passing of her own dreams of stardom, tells *Louise* she sees her achievement not as success but as failure.

Gypsy is a film that she has frequently returned to on video/DVD. Even though she has seen the film many times, she feels *compelled* to watch. Trying to identify what made this film so captivating, the young mother is aware that something about it causes her to *feel uncomfortable*. In particular, there is a confrontational scene between *Gypsy* and her mother towards the end of the film. As the scene approaches she starts to feel uncomfortable. There is an embodied feeling of unease, a slight fluttering in the tummy and a rise in temperature, feelings one might associate with anxiety. The dialogue is a particular cause of this anxiety. *Rose* berates *Louise* and lists all the struggles she has had, scrimping and saving and pushing on to ensure her girls have a career in show business. The theme of the scene is sacrifice and is based on the question "What did I do all this for?" – something the young mother knew only too well from her own difficult relationship with her working-class mother. It is a shock when suddenly the young mother, who is also the author of this book, suddenly realises that her identification had shifted from *Gypsy* to *Rose*. For the first time, I really understood *Rose*. When *Rose* asks, "Why did I do it?" "Where did it get me?", I understood why she felt so frustrated, passionate, angry and why she felt such a sense of loss of her own dreams for recognition.

For the first time I was filled with compassion for *Rose*. My intense familiarity with the scene caused me to question what was happening in the viewing experience. I realised I had a dual and fluid identification with the characters in the scene. I knew how it felt to be a daughter and I understood, for the first time, how it felt to be a mother in relation to her child. The experience of motherhood had changed the relationship I had with the text. It meant something different, and complex new identifications emerged. These new identifications were tied up with my new identity, my past as a working-class woman and my emotional experiences with my closest carers. It was complicated and messy, and somehow *Gypsy* became imaginatively involved in the inner emotional chaos of identity.

* * *

How should emotionally powerful viewing experiences which generate biographical memories, strange or odd identifications and affective responses be understood? Coming from an academic background in film and media studies, I looked to theories for answers. As a film student, I had learnt about film narratives and the role of the camera in creating audience identification with central characters. However, this did not explain why I found this film so difficult, yet pleasurable. Nor why it created such strong affective responses. Feminist theories I had read pointed to the pleasures of texts featuring strong female characters, and

Gypsy was certainly one of those, but I wasn't sure I actually *liked Rose*. Momentarily my identification with her seemed frightening. In a few minutes I had moved through a range of powerful contradictory feelings which I wanted to explain. Then there was psychoanalytic film theory, which proposed the staging of desire. Here was another theory which could not fully explain the strong, ambivalent and often unpleasant feelings the text generated for me. So much of my response to the film was affective and embodied, that it was difficult to put the intense feelings into the specific language or discourse of emotions. There was a kind of amorphous knowledge in what I felt when watching the film. I knew it was something to do with my past but I could not capture or articulate it in language. The categories available in television reception theories of "reading" did not seem adequate to account for *this* viewing experience.

I had been schooled in various film theories. I recalled how the cinematic experience was considered emotionally and affectively powerful, which accounted for its compelling appeal, yet I found myself wondering if the viewing setting of the *home* contributed to the intensity of the experience. I also had a new identity as a mother. I recalled Erik Erickson's (1980) idea that identity change is difficult and anxiety provoking, and whilst I did not fully understand what had happened, I recognised that my new identity had also changed my viewing relationship with *Gypsy*. What, on the surface, appeared like an ordinary experience of watching TV in the home was also very *extraordinary*.

* * *

Themes

This is a book about the everyday activity of watching television and films in the home. The book explores some of the ordinary/extraordinary affective and emotional experiences of audiences of the kind described in the opening vignette. More specifically, it is an account based on an empirical study of viewers' favourite films and television programmes. As I will explain below, the idea of favourites was chosen because this benign quotidian term, in actuality, can be used to help unravel the biographical and emotional intensity of viewing and viewing memories. This may include feelings and emotions that have left an imprint on the viewer. Most people have had the experience of having powerful, often strange and disconcerting, responses to films and television programmes of which they cannot always make sense. This may be a moment, for instance, when they are shocked by the intensity of a powerful cognitive and affective connection with something they encounter on the TV screen in a programme or film. The viewer may note a particularly intense identification with a narrative or plot line which generates a memory, or a strong unexplained affective reaction to what is on the screen, and then this experience becomes inscribed as a viewing memory which they say they will "never forget". This is sometimes captured in a phrase used by some of the viewers featured in this book: "I don't know why, but that one gets me every time." Often there is only a vague awareness of what these responses might signify, and viewers explain that they are often "puzzled" by some of the texts that they like and also the settings and particular times in which they view them.

In part, the book is a phenomenological study which explores the ways in which television and film viewing permeates the everyday life of viewers imaginatively and emotionally, also contributing to the recent interest in emotion, affect and the media. It will also offer insights into the role of memory, identity and viewing. Many of the narratives the viewers discuss in their biographies contain memories of viewing at key moments. This is not, however, a book specifically about the identity and activities of fans and fan communities, although viewers sometimes conflate the terms fan and favourite when talking about programmes and films they view and like. The book takes the position that both words are shorthand for something that has personal meaning and significance. Nor is the book a study of taste, although this idea inevitably emerges in relation to the idea of favourites as cultural choices, and will be discussed in Chapter 1, and in relation to the case studies in Chapters 4–6.

As the opening vignette suggests, the setting in which viewing takes place is also meaningful. Where favourite films are concerned, even when the first viewing of the film was at the cinema, there is something significant about viewing in a domestic setting at home. In part, this is because cinema and television have changed. Time-shift, record-replay and digital technologies have brought film into the home in new ways, and audiences have unprecedented control over their viewing. This means that it is possible for viewers to use technologies to facilitate on-going *relationships* with particular texts and to form personalised viewing styles or user-flows (Caldwell, 2003; Klinger, 2006). I emphasise the word relationships because understanding the nature of the attachments to specific texts, and the creation of accompanying unique viewing practices, is central to this book. The study of viewing settings is nothing new. For instance, ethnographic television researchers have paid attention to the context of viewing, the relationships with other viewers and the television technology (see Morley, 1992; Wood, 2009). What is different here is that I want to add an understanding of the emotional features of the relationships within the viewing and setting and with the text. This is something that has been previously understood, principally, through a sociological lens.

A rich body of research and theory exists in film and media studies offering compelling, if incomplete, explanations of spectatorship and viewing. Although both disciplines were interested in audiences, in practice their approaches were quite distinct and rarely in productive dialogue with one another. Partly, the divisions between film and television studies can be related to what seemed like the irreconcilable differences between the settings of the darkened cinema auditorium and the quotidian domestic location for television viewing. Film theory, for the most part, adopted text-based methods and models that sought to explain the pleasure of the text in cognitive/formalist terms (Bordwell and Thompson, 1979; Thompson, 1988) or through psychoanalytic explanations of the unconscious (Lapsley and Westlake, 1988; Penley, 1988; Mayne, 1993). In these approaches, a universal audience response was read from the text and there was no sense of the spectator as an individual, located historically, and with their own unique biography which might influence the way the text is experienced. Mostly, these paradigms were oriented towards what might be called cognitive or mental processes. Television audience research on the other hand was much more interested in social processes. Approaches can be seen as being divided between either text-reader studies – that covered a range of interests in the text and how it was received and interpreted – or product-user approaches – that focussed on the uses to which audiences put the text and the context of viewing (Wood, 2009). Whilst psychoanalysis and psychological explanations of television audiences were not rejected outright (see Modleski, 1982; Ang, 1996, 2007), it is fair to say that sociological paradigms were highly influential in television audience research. Here, it was the outer social characteristics of the audience, such as class, race and gender, as socially constructed in discourse, that affected the way in which texts were interpreted and used.

Psychoanalytic film theory and sociological audience research both offered explanations about the role of the text in relation to questions of identity and subjectivity. In psychoanalysis, pivotal moments in the development of the infant, such as the Oedipus concept (Freud, 1913), mirror phase or entry into language (see Grosz, 1990; Lapsley and Westlake, 1988) were linked to the cinematic apparatus in accounts of the relationship of the construction of the subject and the fascination and pleasure of cinema. Television researchers explored the way in which television was used in identity work looking, not only at the way the social background of the viewer affected the interpretation of texts, but also the way in which identity could be made and re-made in relation to television (Abu-Lughood, 2002; Wood, 2009; Wood et al., 2008). There was also a strong emphasis on social constructionist accounts of viewing, focussing on the role of discourse and language in the construction of identity. Identification was a key concept in both disciplines as researchers asked "Who or what is it that audiences identify with and what does this mean?" In my viewing experiences I realised that my identifications were complex and multiple; they were inextricably linked to my class and gender identity, the setting, and the qualities of the favourite text. This generated my wish to revisit identity and identification in viewing anew, which will be discussed in later chapters.

There are also two other areas which have not been adequately theorised and which can be revisited through the concept of favourites; pleasures and needs. Implicit in studies of spectatorship, with its language of desire and visual pleasure, is the idea that engagement with the text is pleasurable. Spectators feel compelled to return to the cinema for an experience of sensory visual pleasure (see Lapsley and Westlake, 1988; Mulvey, 1988). Similarly, it has been argued that watching specific television genres, such as soap operas, can provide particular feminine pleasures for predominantly female audiences (Geraghty, 1991). More recently, the idea of pleasure has been called into question. Some writers concur with the idea that, in much research, pleasure is used transparently without paying attention to the specifics of the emotional experience of viewing (Plantanga and Smith in Gorton, 2009, p. 77). Of course, television theorists have also noted that pleasure is only one feature of viewing. Sometimes watching television can be tedious and boring (Tester, 1998) Despite this, in general, going to the cinema and watching a favourite television programme is mostly considered to be something enjoyable. However, as later case study chapters will show, audiences often find viewing unpleasant; in fact they sometimes engage in viewing practices or choose to watch something that makes them feel bad. This raises questions about what we mean by pleasure. Why would audiences choose to sometimes engage with an activity or a text which they acknowledge makes them feel anxious or upset? So, it is time that audience research looks more closely at the emotional nature of viewing and evaluates the idea of pleasure and enjoyment.

Then there is the matter of needs. Here I want to show that it is worth stepping back in time somewhat. Historically, the uses and gratifications paradigm represented a move away from negative views of the "mass" audience, who were viewed in some critical theory traditions as passive and on whom the media had a negative and ideological "effect" (Walkerdine and Blackman, 2001; Arato and Gephard, 1978). The uses and gratifications theory, which took the position that audiences are active and "goal-directed", is now considered to be a rather narrow and outdated approach. Consideration was given to the idea that media and television are used to meet pre-existing social needs, arising as consequences of modernity. For example, the idea that television is a friend to those who are socially isolated and can be used as a form of escapism from the difficulties of everyday life (Blumer and Katz, 1979; Blumer, 1979; Katz, Blumer and Gurevitch, 1974). Accused of behaviourism and functionalism, this approach proposed that television may be used as a form of problem solving, which in McQuail's (1987, p. 73) account, interestingly, points to audiences using television as a "means to *engage with one's own identity, emotional issues and conflict*" (my emphasis). The problem with this proposition is that it is not accompanied by an in-depth account of how this happens or even what it means. However, if the proposition is only partially correct it seems to me that for the individual this is neither insignificant nor purely functional.

Katz et al. (op. cit.) noted that the uses and gratifications theory relies on an underdeveloped view of needs and their origin (which may be in the individual, in the social, or in both) and gets caught up in the idea of "problem solving" without fully explaining what kinds of "problems" the audience is dealing with. Elliott (1974) also deemed this approach too individualistic, not sufficiently sociological and mentalistic, because he saw it as concerned inappropriately with the mental states of individuals. For Elliott, the latter is especially problematic as he believes that is it impossible for researchers to know anything (other than superficially) about the mental state of audiences. It seems that, far from being simplistic, this idea about what audiences do with television emotionally, how they engage with their identity conflicts and, in turn, how television affects them (all of this against the backdrop of a relationship with the social world) actually raises important questions about this commonplace entertainment technology. I think that we might have thrown the baby out with the bathwater where uses and gratifications are concerned (see Chapter 3). I propose that we take seriously the idea of problem solving and conflict and that this can be done by developing an approach which can address this idea of needs *and* pleasures as both social and internal.

In summary then, in the field of audience studies, it seemed the splits between elaborate methods of textual analysis, ethnographic and qualitative empirical audience research, psychoanalytic and social interpretive paradigms, text, context and uses could not be reconciled. However, my viewing experiences seemed to be suggesting that it was time to revisit some enduring and, as I have argued, mistakenly perceived, outdated ideas. It also suggested it was time for a new approach that could take into account inner psychological and outer social processes in the framework of text and context. If a new approach is needed this will also involve developing new methods and interpretive paradigms.

In an attempt to bridge these gaps and create dialogue between disciplines I propose that it is productive to consider audience experience as *psychosocial*, and *Ordinary Audience* develops this concept into a new model for audience research to enhance existing approaches.

Audience experience as *psychosocial*

A psychosocial approach starts from the position that inner and outer worlds cannot be viewed as separate entities. In fact human subjects are the result of both internal and external process which are "co-constitutive; that is, they have effects both ways" (Hollway, 2009, p. 217). This means that psychosocial research can be characterised by an interest in ways in which the "out there" of the social world of discourse, disciplinary practices and power relations is taken up by the "in here" of the psychological and unconscious world (Jefferson, 2008). Then in turn, how the psychic or inner world colours the social realm it encounters and the specific processes involved. Implicit in psychosocial approaches is an acceptance of the idea of a dynamic unconscious, which "presses upon and interferes with everyday life, interactions and relations" (Evans, 2000, p. 13), and which responds in specific ways in terms of unconscious defences (against anxiety for instance) to the world it encounters.

I want to clarify the particular approach I develop in this book and locate it within the wider context of the emergence of psychosocial studies in a number of British universities in recent years. The Psychosocial has not been codified into one distinct perspective or subject area, and an exact definition does not exist. Walkerdine (2008, p. 341) notes that the idea of psychosocial studies is a very British idea, associated predominantly with practitioners from the human and social sciences. The development of the psychosocial has also been characterised by a distinct trajectory marked by interdisciplinarity and the use of psychoanalysis to revisit and shed new light on fundamental issues within social science, for example identity and social change (Clarke et al., 2008). Whilst theoretical differences exist within contemporary psychosocial studies about the degree, if any, of separation between outer and inner worlds (sometimes referred to as psychic reality), most practitioners and theorists would agree that psychosocial research attempts to avoid being reductive to either purely social or psychological factors.

The use of psychoanalysis in the humanities and arts is not new. For example, the ideas of Freud and Lacan have featured in film theory, Frankfurt school critical theory and feminist theory in the post-structuralist vein. The current psychosocial moment can be distinguished, however, by the development of psychoanalytically informed methods for empirical research and a growing interest in ideas from the branch of clinical psychoanalysis known as object relations (see Frosh, 2010, for a detailed account of the development of psychoanalysis "outside of the clinic"). I will explain why this interest has developed below, but before this, it is necessary to take a brief detour to think about this psychosocial proposition in relation to the dominant paradigms in film and television audience research.

Traditionally Media Studies have concentrated on the "social" side of the dyad as a means of understanding inner or subjective processes. For example studies have explored the way language, discourse and "representations" in media texts construct reality and in turn create subject positions, which individuals can take up and occupy. Therefore, most sociological/social constructionist approaches to audience point to this as a one-way process: outer to inner. Conversely, film theory, which embraced Freudian-Lacanian psychoanalysis, prioritised inner processes as I highlighted above.[1] There are only a few studies which pay attention to both sides of the dyad, and this has been further complicated by divisive approaches between text-based and empirical audience research. For instance, Stuart Hall added an important subjective dimension to the encoding and decoding model when he developed the concept of articulation to understand the construction of meaning in relation to texts. Hall recognised that the production of meaning in viewing is not located entirely in the sociological characteristic of the viewer, but emerges out of a process of articulation between the constructed text and the viewer, in historically specific conditions of hierarchy and unequal power relations. Even if "dominant meanings" might have been encoded into a text, subjects bring their own "subjective desires and subjective capacities which enable them to take up positions of identification in relation to its meaning" (Hall, 1999, pp. 309–313). Studies of fandom, though not usually based on psychoanalytic theory, take both text and audience into consideration. Acknowledging the limitations of purely sociological approaches, they ask important questions about the strong emotional attachments to a particular star or genre which characterise the fan identity. Some of these questions are pertinent to this study of favourites. One important question concerns the ways in which the selection of particular figures and forms connects with other aspects of our selves. There is also a supposition, that by making a particular text matter enough to be deemed a favourite, they become a voice for the viewer, which speaks about something that has significant meanings for them (Jenson, 1992; Grossberg, 1992). The need remains to translate these concerns into research design. I will deal with this in Chapter 2. Whilst I have suggested that media studies has not been able to produce truly psychosocial research, this is not entirely the case as there is one notable example to be found in the ground-breaking work of Valerie Walkerdine.

Walkerdine's (1986) seminal essay *Video Replay* stands as a landmark psychosocial study as it took account of the social, historical, biographical and cultural specificity of the working class family. It utilises the psychoanalytic concept of fantasy to show how and why investments of fantasy in the film *Rocky II* (1979, Dir. Sylvester Stallone) could not be divorced from the family and the relational context in which the viewing takes place. She also developed her ideas in a subsequent study on a family's viewing of a video of the film *Annie* (1982, Dir. John Huston) which examined the way television operated as part of a process of relay through which difficult feelings and emotions can be articulated. Again, she found during an empirical study of families with social problems, that aspects of the film's narrative were used discursively to allow the research subjects to speak about difficult familial and social conflicts (Walkerdine, 1997). In both studies by Walkerdine the importance of feelings can be seen as she explores the effect of her presence in the homes of her research subjects and how this played a role in the specific data that was produced and her analysis of this. This is something I explore in Chapters 2–6.

I suggest that both studies can be used as a methodological template for psychosocial research. This is because they are empirical; they take account of the text, the media technology, the social and biographical context and unconscious processes. Whilst the concept of working-class men fighting for their families, which Walkerdine argued in *Video*

Replay, has been widely cited in media theory since, the wider implications for audience research such as researcher reflexivity combined with the use of psychoanalysis in empirical studies surprisingly have not been taken up. Media and film research remains divided, with practitioners doggedly attached to particular theories and approaches, which characterises academic adherence to remain within the disciplinary boundaries in which they work. However, Walkerdine's work demonstrates the potential for psychosocial media research which is reflexive, recognises the way people have affective impact on others and steps out of disciplinary confines, which this book also aims to do. To help clarify this further, it is useful to return to the growing interest in object relations.

Object relations

Object relations theories are often discussed in the context of psychosocial studies as if they are *the* psychoanalytical school of thought of choice. This is not the full picture, and object relations is certainly not the only approach practitioners adopt (Frosh, 2010, p. 203). But, as I hope to show, the ideas of some of the key thinkers associated with this tradition such as Melanie Klein, Donald Winnicott and the contemporary theorist Christopher Bollas can help breathe new life into audience and textual studies. In this book I draw on a range of concepts and ideas from post-Kleinian relations as well as perspectives from classical psychoanalytic thought including the ideas of Sigmund Freud.

At this point, I want to briefly explain why I think object relations has recently become popular and why it is useful for audience studies. Paul Ricoeur (1988) proposed that there is a space where the "world of the text" and the "world of the reader" meet and it is in the interests of researchers to understand what happens in this space. It is my contention that this space is psychosocial and object relations can help us to understand this.

Object relations theories offer accounts of the composition of the self which require the social to be "taken fundamentally into account", but this is not, however, a simple "relationship of cause and effect" as Roger Silverstone (1994, p. 12) has argued. Object relations theories place relationships with objects (which can be people and physical objects or things such as television programmes or films) at the core of human self-experience and development (Gomez, 1997). It is this emphasis, on the relationships formed with the objects that the subjects encounter in the socio-cultural domain, that has made it appealing to researchers. As later case study chapters will explore, whilst human beings might relate to, and put to use, a range of objects, it is important to recognise that human objects – as affective, sensuous, thinking beings – are very different kinds of objects to those objects which are cultural and non-human.

Object relations developed existing ideas in classical psychoanalysis and shifted the focus, from the use of objects as a target for the fulfilment and management of drives as Freud originally proposed, to the quality of our relationships with other objects. Object relations theories offer explanations of internal psychic processes, taking account of the role of the environment in which subjects develop and exist, which media researchers are now recognising offer new ways of understanding the cultural domain. Earliest experiences are said to leave their imprint on the individual, and traces of early life are present in adult experience and interactions with others. This means that individual biographies are significant, but in my psychosocial approach these must always be set against wider historical processes, as later chapters will explain. The way in which object relations theorists conceive of the relationship between the individual and reality has relevance for audience studies, as I will explore in Chapter 1. The infant comes to know and experience reality through its relationship with its primary carer (usually its mother) and is profoundly affected by the nature of this relationship as one of complete dependency from which it emerges as a secure individual. At certain moments in the infant's development the world of objects and culture start to become significant. Winnicott (1971) describes the "transitional space" where physical objects, prototypically the toys of infancy, such as teddy bears, are used by the infant in reality testing. In this space, the infant explores the momentous process of moving from a position of experiencing itself as fused with its mother, through varying degrees of separation to relative independence, where the child recognises itself and its mother as separate beings. The late Roger Silverstone (1994) is notable for bringing object relations theories to bear on the study of television and everyday life. Silverstone took up Winnicott's concepts and ideas, which fall under the category of transitional phenomenon, and proposed that television has all the characteristics of a transitional object (see Chapters 1–3). In part, this means that this everyday entertainment technology is used and related to emotionally, as I shall explore. Winnicott's ideas have also been of interest to cultural and media researchers because he additionally showed how this space was also important in the development of creativity through the idea of play. The processes involved were unique in that they took place in a space that was neither fully inner nor outer. In this space, the borders between inner and outer worlds, self and other, subject and object, are explored, and my argument here is that this is also this space in which the viewing relationships with favourites is played out. I think this concept can help to explain some of the experiences that take place in the space that Ricoeur (op. cit.) observes.

Christopher Bollas is a leading psychoanalytic thinker who links our cultural experiences and use of objects (film and television texts) to psychic well-being and aliveness. He is interested in the way we make choices about which objects/cultural experiences to engage with and argues that our use of objects has an emotional component, which involves connecting with memories and self-states, for example. At the same time, he considers the nature and character of objects, or in this case, film texts, programmes or the activity of watching TV, and asks what it means to be *moved by* an object. This suggests that both the object (text and medium) and the user (audience) are considered jointly. This is especially relevant for favourites, whose characteristics I have proposed include: returning to texts which provoke strong emotional reactions, viewing associated with mood, persistence of memories and repeated viewings. The research covered in this book therefore is informed by the ideas of key contemporary theorists such as Bollas, as well the ideas from the object relations tradition. Until recently, within audience studies, the take up of ideas from object relations has been minimal, with Silverstone's work on television and Mat Hills's (2005) use of Christopher Bollas's ideas to inform his work on fandom being notable exceptions. However, as this volume testifies, the situation is now starting to change. This book can be placed alongside a growing interest in object relations psychoanalysis and culture. The *Media and the Inner World* network in the UK has resulted in a flourish of publications which consider the relationship between the media, culture and unconscious processes. Edited collections such as Bainbridge et al.'s (2014a, 2014b) *Television and Psychoanalysis: Psycho-Cultural Perspectives* and *Media and the Inner World: Psycho-cultural Approaches to Emotion, Media and Popular Culture*[2] apply object relations and other psychoanalytic perspectives to the analysis of culture. Ultimately, this challenges the dominance of sociological approaches by exploring television as an "object of unconscious fantasy, an emotional experience, and a site of psychosocial conflict and struggle" (Bainbridge et al., 2014b).

The focus on relationships is not just about connections between inner and outer worlds. It is also attentive to the intersubjective and dialogic processes involved in our relationships with other subjects. This includes the relationship between researcher and research subject.

It includes dialogic relationships with the medium (TV, Film, gaming, the internet) (see Wood, 2009), and relationships within the imaginary worlds of the text. Chapter 1 will explain these ideas in more depth to show how the home and television viewing create an exceptional psychosocial space.

In some instances readers might encounter the term psychosocial separated by a hyphen (psycho-social). Debate has taken place around the status of the hyphen relating to the distinction and degrees of separation between inner and outer worlds. For some, the hyphen indicates a tension and an important distinction between the two worlds, acknowledging that they are both separate *and* linked. The hyphen attests to the structures and processes that are specific to the workings of the psychic and social worlds (Hoggett, 2008). As Susan Isaacs has argued, the unconscious has its own "dynamic laws and characteristics different from those of the external world" (Isaacs, 1952 in Redman and Whitehouse-Hart, 2008, p. 63, see also Isaacs, 1952). So, the inner world or psychic reality is characterised by its own rules and processes such as splitting, projection, repression (which will be explored later). The outer world is structured, by economic laws for example, and for some the hyphen testifies to a difference which cannot be dissolved (Hoggett, 2008). For others "the boundaries between the inside and outside world are fundamentally blurred and unstable" (Redman, 2009, p. 63), and Frosh (2010) describes the relationship in terms of a Möbius strip where inside and outside flow together. This is a position which refuses the idea of a separate inner world. It has been accused of being social rather than psychosocial in that its focus on the symbolic or language ultimately locates experience and the subject in the social (Hogget, 2008, pp. 380–382; Jefferson, 2008). For Frosh and Baraitser (2008, p. 349) the decision of how to view them as *either* social *or* psychological is "purely tactical". I choose not to adopt the hyphenated version here, but I accept that the relationships and processes identifiable with particular inner/outer domains might be more salient at specific moments and require attention to the specific laws and characteristics of the relevant domain. For instance, when I use the term *identity* I am referring to something that is the consequence of social roles such as class, race and gender and is derived from, and is a consequence of, the relationships people have with social institutions such as schools, work and the state, amongst others. However, identity is also inextricably linked to subjectivity which is viewed as the internalised, subjective and affective experience of those social roles. Subjectivity is, in part, the internal experience of identity and encompasses feelings, affect and emotion in response to our experiences and encounters in the social world in everyday life. Therefore, when I refer to questions of identity throughout this volume I am also referring to subjective processes. This is relevant in relation to the hyphen question, however, as there are moments when outer or inner processes might be more salient and therefore require attention to the laws and processes that are specific to that domain. I also hold the position that psychoanalysis takes seriously the relationship between the two domains and is therefore an important resource to help researchers analyse experience in a non-reductive way, as I will demonstrate in later case study chapters. I want to introduce this further by thinking about some of the concepts and ideas introduced in Chapter 1 and later case study chapters which focus on the way in which inner and outer worlds have an effect on each other.

The psychoanalytic concepts of transference and countertransference refer to innate human capacities present in all our social interactions and describe the passing of emotions and feelings between people, therefore contributing to our understanding of human communication. For Chodorow transference is the "demonstration that our inner world of psychic reality helps to shape, and give meaning to the intersubjective, social, and cultural worlds we inhabit" (1999, p. 13). In classical psychoanalytic theory transference was thought to only be present in the relationship between analyst and client, where the client projects thoughts, desires and feelings which belong to significant figures from their past, prototypically parents, onto the therapist (Grant and Crawley, 2002). There is now widespread agreement that transference is a ubiquitous everyday phenomenon (Andersen and Berk, 1998; Chodorow, 1999; Schore, 2003).[3] This raises questions about the circulation of feelings and thoughts between viewers and the objects that are alive in the world of the text or on TV. As noted earlier, human objects are certainly different to inanimate objects, but films and television programmes feature feeling human beings in a range of real/constructed and fictional texts. This raises a question which I will address in later chapters, about the nature and specificity of viewing and relating to texts as objects with their own special qualities. Previously, the text has been theorised in terms of identification and interpretation, but the recent interest in affect as something "mobile" (Kavka, 2008) suggests it is worth exploring what exactly it is that "moves" between text and audience. This adds the dimension of the movement of feelings between the characters and narrative events in textual objects. This will be explored further in Chapters 1 and 2 and later case study chapters where I consider the home as a psychosocial space and the television as a psychosocial object.

So far, I have concentrated on outlining why psychoanalysis is relevant for a psychosocial approach. I have proposed that the inner world colours the social. Equally, I am interested in the role played by social processes with regard to the cultural choices audiences make. Recent decades have seen social theory concerned with the changes to identity associated with post-industrialisation and the impact of neo-liberalism, both economically and culturally. Identities are now said to be individualised and the successful self is reflexive, disciplined, flexible and self-governing (Giddens, 1991; Beck et al., 1994; Beck, 1992; Beck and Beck-Gernsheim, 2001). Television within neo-liberal capitalism has been theorised as a cultural technology for governance, and it has been shown that the rise of makeover and other reality genres are symptomatic of the importance placed on making the right cultural, or more specifically consumer, choices (see Ouellette and Hay, 2008). In this climate, social divisions based on class are said to be erased to be replaced by discourses of social exclusion, aspiration and self-motivation. The idea of favourites and making choices about what to view, where to view and the style of viewing suggests that questions of cultural choices, or what Bourdieu's (1977, 1984; Bourdieu and Wacquant, 1992) influential work refers to as "taste", have social implications. The success or failure of the self can be judged through the habitus and accumulation and displays of different forms of capital. For Bourdieu cultural practices are involved in the reproduction of social divisions over time. His ideas can be used to understand processes of socialisation involved in the production of a classed self, whose disposition, or habitus, makes the self become recognisable to other group members, in a world of "consensus of meaning and practices" (Bourdieu, 1977, p. 80).[4]

Taste not only displays our class location, it also translates into forms of economic and symbolic capital. One effect being that to be recognised as working class is particularly undesirable (Bourdieu, 1984; Skeggs, 2005). Bourdieu does not theorise these processes as emotional. However, it seems obvious that being mis-recognised or not performing the classed habitus and dispositions correctly can provoke anxiety and fear. Equally, if group and class distinctions are becoming erased in favour of the individual, then the performance of individuality could potentially generate worries and concerns about success and failure. It is this development on which recent theorists of reality TV have focussed. In particular, noting a discursive link underpinned by psychosocial assumptions between the successful production of an "outer" self, through training in the correct way to consume, which is then linked to "inner" happiness and fulfilment (Ouellette and Hay, 2008; Whitehouse-Hart, 2014). We know very little about how these processes of class, taste, habitus and individuality might actually be experienced emotionally, particularly in the private viewing acts. Do these social demands have any effect when viewers chose to watch something? How do viewers feel about their tastes and cultural viewing practices? Likewise, do viewers' tastes and viewing practices make viewers *feel* and if so in what ways? Finally, and importantly, how can we research these kinds of questions and experiences? At this point, then, it will be useful to introduce the study and the method.

Meet the viewers

Chapter 3 explains that it is not possible to produce psychosocial knowledge whilst working within the confines of established research practices. The insights of psychoanalysis force us to question our perception of reality. The idea of transference, for instance, raises questions about forms of unconscious intersubjective communication between the researcher and research subject which shape the production of data and its interpretation. Theories of the dynamic unconscious force us to question the idea of objectivity and self-reflexivity in research. The inner world is at "work" when we do research. Therefore a psychoanalytically informed research design was adopted. The purpose of this was two-fold: to assess the use of psychoanalysis in empirical audience research and to develop a psychosocial analytical framework for audience studies. This means a form of analysis that pays attention to social and psychic processes, and it achieves this by delving into the world of audience experiences, memories and attachments, through the idea of *favourites*. This productive concept is able not only to address the biographical significance for viewers, but also to provide a way into the emotional features of viewing. It allows for greater understanding the process of textual interpretation within the context and tensions associated with the moment and the historically specific viewing context.

The study on which the later chapters are based was not designed to produce findings that can be generalised. Given that audiences are so diverse, I think a generalisable approach is implausible. My study is designed to offer new theoretical and conceptual insights for media audience research around experience, identity and emotion in viewing. Traditional questioning techniques associated with qualitative interviewing were reviewed and found to have limitations in relation to the psychosocial questions and approaches I wanted to explore. With these shortcomings and my research aims in mind the resulting study was based on a series of long, in-depth interviews and letters with a small group of viewers, influenced by narrative interview methods; Biographical Narrative Interview Method (BNIM) (see Wengraf, 2001) and its development in a modified version, Free Association Narrative Interview method (FANI) (See Hollway and Jefferson, 2000). These methods are psychoanalytic and socio-biographic, and they are designed to generate narratives in interviews which are assumed to be expressive of both conscious *and* unconscious concerns, which also bear the imprint of individual and social processes. So this is a study whose method is concerned with the production of data through emotionally significant stories and memories. It reflects not only on the narratives that interviewees offer but also on the research process involving psychoanalytically informed methods. As some of these ideas and approaches may not be familiar to some readers, it is important to explain some of the epistemological assumptions and methodological considerations associated with the use of psychoanalytically informed methods, which will be covered in Chapter 2. In all research with human subjects, there are inevitable ethical questions which will also be touched on in this chapter. Chapter 6 however will examine some of the ethical tensions "in context" through a case study.

At this point, I want to pause and introduce the reader to the viewers who took part in the research and who feature in the case studies in Chapters 3–6. I have changed names and biographical details to maintain anonymity[5]. I am grateful to those who took part for their permission and willingness to share their experiences and thoughts. It was also pleasing to find that those who took part also expressed that they enjoyed talking about those puzzling experiences and questions about their likes and dislikes which also fascinated them.

Mary (48), Sue (43), Daniel (37), Sophie and Chris (both 29) and Bill (65 retired) are examples of what I am calling ordinary audience members. This is partly ironic and points to the nature of watching television and films as something quotidian and unexceptional, whilst involving complex psychosocial processes of identity and interpretation. Viewing is, and viewers are, both ordinary and extraordinary. In part, the term ordinary is also being used to refer to the kind of audience member that has been seen in previous audience research – working-class women and men. The study represents a snapshot of the viewing practices and tastes of a group of people who were born into working-class homes. What distinguishes them is that they have each witnessed the emergence of technological changes, set against a background of historical and social changes, which have changed the nature of cinema-going and also television viewing in the home. Recent years have seen the emergence of the term post-cinematic era. This does not mean that cinema-going is dead or irrelevant, but it does point to a change over time, which has seen film viewing increasingly take place in the home on a television set (Klinger, 2006). As I explained above, as well as this shift away from the cinema to the home for film viewing, audiences have also witnessed changes in television viewing, such as time shift and the development of digital technologies, which have radically altered their own viewing experience and use of television. This book examines the impact of these changes on viewing and textual identification. Each viewer grew up in the era of public service broadcast dominated television, where they lived in a house with only one television set and where the family had to negotiate viewing around a choice of three or four channels, which was largely dictated by the broadcaster's schedule. The transformations in broadcasting also took place against the backdrop of social changes associated with late modernity and globalisation, for instance increasing affluence (Goldthorpe, 1969), growth of consumption, social mobility, the emergence of "risk" society (Beck, 1992), "self-reflexivity and life politics" (Giddens, 1991) and increasing "individualisation" (Beck and Beck-Gernsheim, 2001).

The sample chosen for the research represents a spread across the age and gender spectrum of white British working-class people, who have frequently been the subject of cultural studies and media research on audiences. These people occupy various positions in relation to social and cultural changes, such as upward and other forms of social mobility. This combination allows us to revisit established research territory anew. In particular, to explore the emotional costs associated with social mobility and whether a study of viewing can provide a new but effective lens on the experience of social change.

The study was not designed as a comparative study to oppose middle-class and working-class experiences and responses. Instead, it used a small sample of people with differing positions on the age spectrum; from mid-twenties, through middle age (40s) to retirement, to provide different perspectives on the socio-historical and technological developments I have described. The research concentrates on the lived experience of class and gender, assessing emotional costs and the way gendered and classed lived experience influences particular forms of attachments with texts. All viewers featured began their lives as traditionally working class, coming from families where parents were manual labourers and where money was often tight. At the time of the interviews, the female participants were mostly working part-time and the men full-time or recently retired in a range of occupations from care work, clerical and administration to professional employment in education. All the participants were over 25, which was relevant as the research questions were interested in relationships with texts that had developed over time, as well as the impact of biography on interpretation. It was decided that mature participants were more likely to have the kinds of relationships with texts and viewing that the research wanted to explore.

The demographic is important in contextualising the viewers, but I wanted to give the reader a flavour of who they are and what kinds of enigmas and puzzles their viewing sets up: what is interesting and complex about these viewers, their lives and viewing. With the exception of Bill, the viewers had their strongest emotional reactions to films viewed in the home. Whilst each case is unique, providing idiosyncratic findings, important conceptual themes emerge which connect these cases together and proffer ways of understanding the viewing experience and relationship with texts. As the case-studies presented in later chapters show, biographical events do not stand outside the viewing experience, but they shape identification and patterns of fantasy and interpretation of texts. I want to briefly capture some key moments from the case studies to set the scene before moving on to a more formal discussion of the central themes and debates covered in the book.

Daniel and Chris

Daniel and Chris, who we meet in Chapter 3, are young men each living alone in their own home, working and building careers. For both, watching television at home is an important feature of their leisure time but one that also generates anxieties. Daniel can always find something to get "lost" or "absorbed" in. He paints a picture of his life as structured around the polar opposites of work and free time. His free time is filled with watching television programmes and films. As he explains, "the thought of a good programme or film at the end of the day makes it worth going to work". After "you've tortured yourself for that many hours" (at work) he plans to "make the most of his viewing time". An avid viewer, Daniel talks enthusiastically about things he has viewed and types of programmes he enjoys. He has numerous "favourites" he has returned to "ten times or more". Daniel takes advantage of time-shift technologies such as his Sky Plus to ensure he has programmes lined up to fill up his free time. He prefers to watch an entire series in one day rather than an episode a week. Viewing for long, intense periods began when he was nineteen. For the first time he stayed at home alone while his parents went on holiday. He spent the entire week "absorbed in watching *Twin Peaks*" (1990, Dir. David Lynch, BBC 2). This chapter explores the meanings and pleasures of intensive and absorbing viewing practices.

Chris is single and has his own home, where he lives alone. He likes "classic", "cult" and "low budget" films, and he is a viewer who appreciates cinema as a "superior" experience. Despite a taste in film that might suggest he is a cineaste, like others interviewed he prefers to watch films at home. However, watching films at home is not always a *pleasant* experience, partly because he "finds it difficult to relax". He explains he "gets twitchy". Chris would like to watch television or films to help him "switch off" but instead finds he continually feels like he should be "doing something", including mundane chores and housekeeping or more challenging intellectual study. Many of the films he views would be considered intellectually demanding, and often he feels "threatened" when he "doesn't intellectually understand" a film he is watching. In response to this he disciplines himself, making sure he "reads around" to "make himself understand it". In contrast to these disciplined and uneasy viewings Chris has found a television text, the US sitcom *Frasier*, which he finds very "soothing". He has collected videos of the entire series and plays episodes most evenings in his bedroom to help him relax. This raises some interesting questions about viewing and anxiety. The chapter explores these questions specifically in relation to questions of taste, masculinity and time.

Sophie

Recently married Sophie, the subject of Chapter 4, describes herself as someone who forms strong emotional attachments to what she calls "things" and "stuff". She is a Disneyphile and has, what appears to be a very distinct fan identity. Her flat is crammed with Disney merchandise and memorabilia as well as items with a 1950s America theme. These are important imaginative spaces for Sophie as they capture some favourite films which she repeatedly watches at home: *Back to the Future* (1985, Dir. Robert Zemekis, USA) and *Toy Story II* (1991, Dir. John Lasseter, USA). Sophie recreates the feelings associated with Disney movies through her memorabilia. For Sophie, however, her "things" are not just "stuff". Like the toys in the *Toy Story* movies she imagines they have thoughts and feelings, coming alive when humans disappear. When cowboy doll *Jessie* is put in a clear-out box for a charity shop in *Toy Story II*, Sophie cries uncontrollably as she identifies with her own perceived abandonment as a child when her father left the family to live with another partner.

Back to the Future's narrative and mise-en-scene of the 1950s represents a "carefree space" which Sophie seeks to recreate in her home through the acquisition of "things". Sophie is also someone who uses her DVD player to engage with her favourite films. In fact, *The Wizard of Oz* became a favourite film in her twenties after she purchased it on video. Through this repeated viewing she very quickly formed a strong attachment to the text, which is about a young girl alone who makes friends to enable her to survive, which Sophie strongly identifies with. However, she is often puzzled by the selections she makes. She explains she frequently chooses texts (often films) for viewing that she knows will "make her feel bad". In fact she explains, "I don't know why I watch it, it makes me feel worse." Sophie's study is interesting as it suggests a number of questions about mood, memory and the idea of pleasure as something that can be called *dis-pleasure* as an example of the contradictory features of viewing experiences. This will be explored through and exploration of the role played by consumption and viewing in relation to psychoanalytic understandings of mourning and repetition.

Sue and Mary

Chapter 5 introduces Sue and Mary. Both are in their mid-to late forties, working class, married and mothers to sons, with whom they have intense relationships associated with strong emotions of love, worry, adoration and fear of loss. Sue has a secret which she shares in her interviews: she has a passionate attachment to an unconventional sex symbol, the English actor Christopher Eccleston. The intense relationship began with the acclaimed television drama *Our Friends in the North* (1995 BBC). Following this, she went on to see "everything he has been in" and likes the fact he is not conventionally attractive, as that makes her feel he is "exclusively" hers. Sue describes watching films at the cinema as qualitatively different to watching films at home on television because at home the experience is often emotionally intense. For Sue this is a "puzzling" experience which she would like to understand. Sue describes one such powerful moment from the film *Big* (1988, Dir. Penny Marshall). This was not only a moment to which Sue had a strong emotional response but, also, the moment is significant in terms of her identity as she says she feels that the scene also "sums" her character up. Sue finds the emotional responses she has had at home "fascinating". She explains there is a moment (and it is just that, a moment), when Tom Hanks' character *Josh*, a child turned by magic into an adult, returns home to his mother. He is left wearing his executive suit, now far too big for him, as he walks back into the family home. She is "not sure why" but she finds herself "filling up just talking about it" and it makes her cry "every time" she sees it. "It's something to do with his mother and the big suit" she says. She asks a question many viewers ask about texts they have had emotional responses to: "I don't know why but that one gets me every time?" I will offer some reasons why these texts and attachments are so intense for Sue in this chapter.

Chapter 5 also introduces Mary who explains that it was as a teenager that her "style" of viewing, which she calls "layers", began. "Layers" is a complex response to the difficulties she encountered around her gender identity in her teenage years that were also unique to her specific family composition. Viewing in "layers" involves a vibrant imaginative and fantasy life which involved "going beyond the characters" and imagining their psychological motivation in light of her experiences. For Mary, prior to becoming a teenager, she recalls watching television as an important part of her childhood but one she associates with being "scared of things" she saw on TV, which she recounts in her interviews. One evening, as a small child who "used to do a lot of wandering in the night", she walked in on her parents who were viewing *The Birds* (1963, Dir. Alfred Hitchcock). She was terrified at the sight of a woman having her eyes pecked out. The incident gave her "nightmares for years and years" and "to this day" the memory still haunts her. This chapter will explore the reasons why this incident was discussed at length in the interviews about favourites and suggests psychological processes involved in viewing in "layers". This chapter draws on the ideas of Christopher Bollas and introduces a range of forms of identification present in in the object selection of particular film texts; idiomatic, biographical, sibling and intergenerational. It explores those powerful memories that viewers find they do not understand but are unable to forget.

Bill

Bill (Chapter 6) is a highly educated retired man who prefers the cinematic experience to watching TV. He is also a cineaste with an intellectual approach to films. Bill is knowledgeable about cinema and film, reading texts like a film scholar and he is *au fait* with the language of film theory. For Bill, who has entered the latter part of his life, the discussion of favourites was an opportunity for reflection and memory, which he relished. Whilst he discussed specific film texts, his reminiscences were mostly directed towards the viewing setting and the biographical events that formed a backdrop to his favourite or other memorable films. Bill was interested in taking part in a project that explored favourite films as he saw it as an opportunity to engage in a process he called "knowing thyself". He wanted to talk about films to "find out more" about himself, a process that he saw as both "exciting and dangerous". This chapter investigates what might be risky and dangerous about discussing favourite films and explores the relationship between researcher and interviewee to shed some light on this. This also includes a discussion of some of the ethical dimensions of research. The chapter also reflects on viewing memories and biography at the latter stages of the life cycle.

Chapter 7 is the concluding chapter which draws together the themes from the theoretical, methodological and case study chapters. The chapter provides a review of the new concepts and ideas the book introduces and reflects on the appropriateness of psychoanalytic concepts for audience research. It draws on the rich, creative, powerfully affective and emotionally laden experiences of viewing, as well as the distracted, routine, difficult and boring, proposing that we understand these various often puzzling facets of audience experience as examples of psychosocial processes. This illustrates the importance of understanding personal meaning in understanding the reception of texts and the uses and contexts of viewing. Also arguing that it is essential to make links between the biographical experiences of audiences, history and the impact of wider social processes which show how viewing sits at the interface of both psychic and social worlds.

As the reader can see from the introductory outlines, each case study offers scope to explore a range of emotional and social processes in which viewing is involved. In the light of this *Ordinary Audience* has a two-fold purpose: to introduce a range of psychosocial perspectives and methodological approaches to a media studies audience and to develop the psychosocial intellect landscape through the exciting addition of audience and popular culture research. *Ordinary Audience* will bring psychosocial interpretive paradigms and psychoanalytically informed methods to bear on these case studies. It is my assertion that in-depth immersion into the world of viewers is the best way to begin the process of psychoanalytic theoretical and conceptual reconsideration of audience studies to life.

1
Favourites, TV and Home: Psychosocial Perspectives

The previous chapter explained that this book will develop a new approach to audience studies and that this approach is distinctive in that it is psychosocial and psychoanalytically informed. I began the book with the assumption that all experience is psychosocial. However, in this chapter I am going to explore the specific psychosocial essence of favourites and show how watching television at home has intense psychosocial resonance. The opening pages also contained the assertion that psychoanalysis, and object relations theories in particular, can offer a fresh lens to illuminate the specific nature of psychosocial experience. Therefore, this chapter will explore some theoretical perspectives from psychoanalysis to illuminate the following themes, which also constitute the sections in the chapter, namely: favourites, within the context of object relations; biography and emotional experiences; the effect of technological developments in entertainment technology on attachments and relationships; and television and the home as affective psychosocial objects and spaces. I will begin with favourites.

Favourites and texts that leave an affective impression

Mary (see Chapter 5) was sitting at home one day watching on television what she classed as a harmless, bland film: the comedy *Father of The Bride* (1991, Dir. Charles Shyler). There was a moment in the narrative where the humour was created around the father's failed attempts to capture a moment alone with his daughter, to say goodbye before she left for her honeymoon. As she watched, Mary was shocked to find herself, in an instant, overcome with emotion; her eyes filled with tears and she began uncontrollably sobbing. This was an experience she will "never forget".

Annette Kuhn (1995, p. 33), motivated by her own similar, unexpected and highly-charged, emotional response to a screening of the film *Mandy* (1952, Dir. Alexander MacKendrick), recognised that emotional responses in film and television research had been neglected. She was not able to find a theoretical paradigm to explain her experience, and all she knew was that her response came from the past and that it was *troubling and impossible to ignore* [my emphasis]. Emotions, she argued, cannot be used uncritically as some kind of a guarantee or evidence of authenticity but, at the same time, emotional responses cannot simply be discounted.

Most viewers have had the experience of going to the cinema and seeing a sad film or watching a television programme and they may have struggled to fight back tears. Crying and feeling sad, however, are not the only emotional responses viewers have to texts. Mainstream films and certain television genres such as drama, soap opera, reality and talk television produce texts that are designed to elicit various emotional responses from tears, anger and disdain, to delight and joy in the viewer; it is part of their function. This also operates ideologically, but I shall not deal with this issue here. These trends have been observed and documented in research (Walkderdine, 1986, 1997; Ang, 1985; Mumford, 1995; Smith, 1995) with many theorists placing emphasis on the discursive features of emotion on television and how audiences articulate their responses in interviews (see Tolson, 2011; Wood, 2009). Until recently, few attempts have been made to systematically theorise the different ways and processes by which emotion might happen for viewers.

Prior to the recent interest in affect and emotion in television viewing (see Gorton, 2009; Kavka, 2008), little effort had been made to theorise emotional responses when viewing (Kuhn, op. cit.). Much theory existed that theorised pleasure in the cinema and, as I have outlined previously, this often glossed over the emotional content of cinematic pleasure. Notable exceptions have been work on melodrama in film and television (see Neale, 1986; Ang, 1985). Steve Neale (1986), for instance, offered a plausible account of why spectators cry at melodrama. He argued that this must be understood generically as a function of narrative organisation within the text, therefore locating the emotional process as a function of the text.[1] Ien Ang (1985) and Harper and Porter (1996) both based their accounts on engagement with audiences. Ang's seminal study of *Dallas* argued that the pleasures of the soap were based on audiences' emotional engagement with the programme. Viewers identified with the emotional realism of its melodramatic imagination and tragic structure of feeling and the *pleasure* revolved around being swept away by emotion, showing viewers actively seeking out emotion. Harper and Porter's (1996) study of the questions asked about crying at movies, in the mass observation questionnaires of the 1950s, identified particular emotive scenarios which moved people to tears. Often these responses displayed gendered features, such as partings, unhappy children or cruelty to animals and death scenes for women, and patriotism, self-sacrifice and heroism for men (1996, p. 157). This suggests culture and gender to be significant factors in the production of emotion in viewers, and we must bear in mind that this is also a historically specific study and concerns displays of emotion in the public sphere. I will return to this later.

Staying, for the moment, with the idea of emotion and crying, it can be said that one approach to viewing and emotion might be seen as a seeking out of emotion because it serves possible cathartic functions both culturally and individually (O'Shaughnessy, 1994, 1999; Chaudri, 2001; Gibbons, 1999; Hamilton, 2001). Harper and Porter also found that answers to questions in the mass observation study suggested cathartic responses to films (1996, p. 153). Equally, in a uses and gratifications vein, films have also been cited as vehicles for "problem-Solving" (Berg-Cross et al., 1990; Peske and West, 2002)[2]. Audiences are active and they use texts, this is established, but we need to know more about what exactly they are using them for and what this might mean emotionally.

Clearly, seeking out an emotional experience is one feature of viewing, and later chapters will show that viewers, influenced by mood, do sometimes select a favourite text to connect with a particular emotion or feeling, even if sometimes the viewer is not clear why they seek to do this or what it achieves (see Chapters 3–6). Here I want to note that most existing theories do not fully account for the unexpected intensity of the kind of emotional experiences described above, which become one kind of unforgettable favourite for audiences. I want to think more about the nature of what viewers discuss when they are asked to talk about favourites. Therefore, the first feature of favourites that I have identified is what I want to call an encounter with an *Impact* text.

Definitions of the word impact include: collision, the effect or impression of one thing on another and the power of making a strong immediate impression (Webster's Dictionary). These phrases capture some of the significant viewing experiences this book is concerned with, such as the shock and unexpectedly strong reaction viewers can have to a text. Often what impacts on the viewer, as illustrated in the above example, are mere *moments* from a film or programme, perhaps a visual image or a particular instance of plot development. This moment is often stored as a memory and is distinguished by an association with an ill-defined, unexamined and often surprising emotional impact at the time of viewing. Often these instances, like the example described by Mary above, do not conform to the moment in the text where emotion is generically or narratively encoded into the film/programme as suggested in some theories (see Neale, op. cit. or Ang, op. cit.). For viewers, they remain unforgettable experiences. Kuhn (op. cit.) argues that emotions and memory force experience onto the agenda. The next section explores some of the ways this happens and in turn this helps to bridge the gap between text and audience response.

Impact texts: biography, memory, attachments and repetition

I propose that there are some other distinguishing features of impact texts and favourites. The first of these is that favourites and impact texts have *biographical emotional* significance for the viewer. This can be approached through the related areas of memory and biography. The second involves *attachments* to objects and the notion that certain texts and also viewing practices are *returned* to and repeated. In both cases, emotion is woven inextricably into the fabric of biographical memories and attachments. Harper and Porter's (1996, p. 153) research captures some of this. They found when answering the question about crying at the cinema, that "memories of key scenes from films gave respondents a kind of fast-track access to emotional high points in their lives". They noted that the respondents did not primarily answer the questions in response to the emotional intensity of what they encountered in the text, its form or narrative for instance, but in response to its personal biographical and emotional significance. Whilst the research on which this book is based did not ask interviewees questions about specific emotional responses (see later chapters), a central finding was that emotion, not just crying, but also frustration, boredom, joy and anxiety, as well as memory and mood, emerged when favourites were discussed, and this had biographical significance.

Work in the memory tradition by key figures such as Frederick Bartlett (1932) suggests that memory is associated with attempts to impose meaning on the world, as well as being evidence of the influence of the "world on the individual". However, this suggests a rather one-way process, with the memory being the end process or effect of the social world. The memory produced in the telling is also a product of each person's past experiences, their current mood, and a range of other expectations that are both specific to the moment and the past event (Foster, 2009, p. 13). This also suggests psychological and psychoanalytic understandings. When watching films and programmes the individual brings their life experiences to the text, and these inevitably involve psychic processes. The memory event, in this case a particular screening/viewing and the textual characteristics, are one part of what is remembered. As later chapters explore, there are a range of other internal and external processes taking place as well.

Grossberg (1992), writing in fan studies, has also suggested that it is important to consider biography to understand what might be happening emotionally for audiences when they encounter their favourite texts. He concurs with the idea that audiences are meaning-making beings and that they constantly struggle to make the text mean something that connects specifically "to their own lives, experiences, needs and desires" (1992, p. 52). Texts, which are special to the viewer, will have a particular feeling and foreground a certain emotional sensibility, which chimes with the individual's particular circumstances. Grossberg (1992, p. 57) argues that by making a particular text matter enough to be a favourite the text becomes a voice for the viewer, speaking about something that has significant meanings for them, although a psychoanalytic corrective to this would note that the meaning might be unconscious, misunderstood or misrecognised.

Many biographical and narrative research paradigms are influenced by theories of discourse and social construction and suggest that narratives are important in the construction of the self (Roberts, 2002; Polkinghorne, 1998; McLeod, 1997; Freeman, 1993, see also Chapter 2). For narrative theorists then, symbolic materials (in this case texts) are used in projects of the self and woven into narratives which are told to the self and in the case of research to the interviewer (Freeman, 1989, 1993; Thompson, 1995). Thus, in many of the studies that include memory and biography and meaning, it is assumed that the narratives constructed are involved in processes of the self or identity work. Biographical narratives, which involve viewing, share some of the features I have outlined: the strength of the affective response; memories with biographical significance; attachment to the text; the way in which the memory forces itself into the telling of the interview situation and the element of puzzlement. In the case of the latter, many research narratives begin or end with a statement about the interviewee's bemusement such: "I don't know why I watch it when it just makes me feel bad?"; "I don't know why I like it?"; "I don't know why I'm telling you this...?"; "I don't know why this has come to mind...?" These statements suggest that meaning is not fully achieved as there are aspects of the viewing process and its reiteration in narrative interviews that defy logic and meaning somewhat. It is this puzzling aspect that psychoanalysis can help to unravel as I will explore presently.

Turning now to the question of attachment, it is useful to explore some existing approaches as well as qualifying what this means in relation to favourites. Attachment to watching television has been seen as a social problem, with the figure of the couch potato becoming synonymous with social decline as a consequence of modernity. Television has been perceived negatively in terms of its social effects captured in Postman's (1985) famous concern that people were amusing themselves to death (see Chapter 3). In a uses and gratifications vein, attachments to watching TV were also theorised as consequences of social change in post-war industrial societies. Here, television was associated with forming what Horton and Whol (1956) called "para-social" or "surrogate relationships", which replace face-to-face interaction, with television being an intimate medium. Chapter 3 considers this when it contemplates the emotional consequences resulting from societal change which sees individuals needing to use television as a friend.

Fan studies have also considered emotional attachments (and associated practices) to texts, stars and genres, and the idea of pleasure has also been important in this field: there is something pleasurable and fulfilling about being a fan. The word fan is on one hand a common word used laconically for general tastes and likes, linking it to discourses of taste, which will be addressed later. On the other hand, it has been conceptualised as an emotional attachment and a relationship which includes "cherishing, desiring, admiring and celebrating" (Jenson, 1992, p. 25) the object of affection. It has also been described as a particular form of loyalty, an attachment with active and affective elements, which Jenson called "enacted affinity" (1992, p. 19). This implies, for instance, the collection of memorabilia, writing fan letters, or attendance at awards ceremonies or exhibitions (Harrington and Bielby, 1995). Some of the viewers featured in later chapters describe themselves as fans. For example, Sophie would fall into the enacted affinity category, as she is a Disneyphile and collects and consumes film and television memorabilia (see Chapter 4). Other interviewees, Daniel, Chris, Mary and Sue (see Chapters 3 and 5), use the term fan in the shorthand way to capture their attachment to a particular text or star without the enactment of overt fan identity or behaviour. In relation to fandom, Jensen (justifiably) contends that we do not know enough about the *nature* of attachments and investments that are manifested in people's lives. For Jensen, like others, the question is about meaning. She wants to understand how our "selection of particular figures and forms connects with other aspects of ourselves?" and, if so, "What do our favourites mean?" and "How do they mean?" (1992, p. 25). In the years since Jensen made this claim I would contend that we have not made any great leaps in audience studies that would enable us to confidently state that we understand favourite attachments and their meanings. My argument in this book is that this is in part due to the lack of a fully developed psychosocial model for research and also because sociological audience researchers have refused to accept the insights from psychoanalysis, in particular the idea of unconscious meanings, as I will show in the next sections when I introduce a psychoanalytic perspective.

Attachments are not just emotional or intellectual connections (see Bill, Chapter 6) made to particular texts, genres or stars, although this is one feature of favourites that I have explored. Attachments are also made to particular practices. For instance, in Chapter 3 we meet two viewers who have favourite texts that they like to view in particular ways. We will see that Chris likes to watch *Frasier* in his bedroom at bedtime. Daniel prefers a certain "style" of viewing involving "saving up" programmes/films to watch in "large chunks", and whilst this is a generalised practice, it is also something that developed in the 1990s in relation to his favourite television series *Twin Peaks* (1990, BBC, Dir. David Lynch). In both cases this brings in notions of the text, context, mood and time (see Chapter 3). Attendant to this notion of viewing style is the idea of returning, revisiting and repeated viewing. I have suggested that attachments and favourites can manifest in terms of unforgettable memories about texts and viewing practices, often that audiences ponder about, without reaching any convincing conclusions about their significance or meaning. As later chapters will show, sometimes viewers find a compelling desire to return to a particular text they might have seen many times. Equally, replay technology enables a re-kindling of a relationship with a text not viewed for many years. Perhaps a text originally encountered in the viewer's past is selected and revisited, becoming a favourite only on the revisiting. This happened to Sophie (Chapter 4) when she found herself, as an adult, drawn to a film she had seen as a child (*The Wizard of Oz*, 1939, Dir. Victor Fleming), which then became one of her most watched and revisited favourites. Her relationship with this text continues to develop over time and has become associated with enacted affinity in terms of a large collection of memorabilia. If favourites are also characterised by repetition, then this has come about and been intensified by developments in viewing technology.

As noted in the introduction, television and film have changed, and continue to change, with the development of home-viewing and digital technologies. Record-replay and time-shift technology such as VHS video and DVD, along with computers and home-viewing cinema systems, have replaced cinema-going as the main mode for viewing films and television. Increasingly, computers, tablets and even mobile phones are becoming the device of choice for watching television and films. The impact of the VCR on the micro-politics of family viewing has been discussed extensively, and these developments are not a central concern of this book so I will not discuss them in depth here (see Morley, 1992; Gray, 1992). What I am concerned with is the fact that the digital landscape and home entertainment record-replay and time-shift technologies have something in common; they allow viewers to *personalise* their viewing in previously unimaginable ways.

Caldwell (2003) challenges Raymond Williams' (1974) description of the institutional organisation of broadcasting as "flow", where the experience of viewing is "dream-like", as viewers watch TV rather than individual texts. Caldwell argues that power is no longer fully in the hands of broadcasters to direct viewing, as now audiences navigate from television to the Internet and across numerous platforms and channels in the creation of their unique viewing user-flows. These user-flows will inevitably include a mix of television, film, music videos and games. In relation to films, in the past spectators would have to see a film at the cinema whilst on general release, with the invention of television acting as a rather unreliable medium for catching a second viewing of a film previously enjoyed. Invariably, some films that become favourites have only ever been seen on television as described earlier in the case of *Gypsy*. Films have always been part of the flow of watching television which Williams (op. cit.) has described. Nowadays, it is not always necessary to catch films at the cinema when released, as viewers do not have long to wait before the film is available on DVD and movie channels. Many films spectators have enjoyed, even if viewed for the first time at the cinema, now find their way into the home either through a television screening or on DVD or other digital film/television channels. In the case of television, it is possible to buy classic television series on DVD, and there are numerous opportunities to catch favourite series and films again on dedicated digital channels. Even if viewers miss a programme they like, they only have to push a plus-one channel or on-demand option, making the idea of appointment television not irrelevant, but certainly less significant than in the past.

Established television and film theory falter in the light of these developments. These changes make the idea of active and passive audiences redundant, as viewers now have the chance to build unique relationships with texts and also develop personalised viewing styles. As I explain in Chapter 3, in the example of Daniel who likes block viewing technologies such as *Sky Plus*, these technologies allow viewers to record an entire series at the touch of a button, where previously video recorders could only be programmed for one or two programmes. Significantly, the technology can recognise and record programmes and genres that viewers like. Developing this further in relation to favourites, I want to explore the idea of repeated viewings and returning to texts for subsequent viewings as a consequence of technological developments.

Bourdieu (1984) used the term "habitus" to refer to the enduring habits, routines, engagement with particular forms of culture and cultural practices that people come to associate with their disposition or character. For Bourdieu, culture is a social agent that has effects, as it is involved in the demarcation of social class groups, most notably through processes involving the demonstration of the possession of taste. For instance, engagement with the different forms of culture associated with legitimate (high) and popular culture corresponds to social class distinctions. Whilst our use and engagement of culture may appear natural, it is a result of socialisation processes in which agencies such as education and, most importantly, the family train the individual to understand and appropriate legitimate forms of culture into the dispositions and routines of everyday life. This use of culture translates into capital and confers differential social power on social class groups and the individuals within these groups. Following this, we can think about the way taste and the texts that we return to might relate to such social processes. Interestingly, Bourdieu notes that the end result of this form of socialisation is that our habitus and tastes become unconscious and that we unconsciously "bear the traces of the past in our present selves" (1977, p. 79). Inevitably, this points to the effects of the unique events and experiences that occur biographically in families even in this broad theory of social class socialisation. I will explore these unconscious aspects of taste carrying the traces of the past and also processes of tastes as anxiety provoking in Chapter 3. In contrast to this idea of socialisation becoming unconscious, recent social theory has suggested that contemporary society is "ego-centred'" (Beck, 1992, p. 136). Modernity is said to be "reflexive" and the ego performs reflexive, decision-making functions.

A characteristic of this (which is linked to a growth in consumption), finds cultural products used to define the individual as an *individual*, in a climate of increasing individualisation (see Beck et al., 1994; Beck and Beck-Gernsheim, 2001; Beck, 1992; Giddens, 1991; Featherstone, 2007). In Bourdieu's account, the unconscious is evoked without an explanatory account of the specific unconscious processes or the role of the inner world. In Beck's account, the ego reigns supreme, as if modernity and consumer capitalism has overcome the irrationality usually associated with the unconscious. My aim here is not to debate these theories, but to note that for both, questions of taste, consumption and object are implied. Both theories draw attention to the idea of motivation, object selection and habit but with no sense of how unconscious processes might be involved. Despite evoking psychological language, both accounts do not offer an adequate explanation of psychosocial processes involved, nor any recognition that taste, reflexivity and trying to be an individual might be lived as emotionally difficult experiences. They also cannot account for the intense emotional experiences audiences have in relation to favourites and impact texts. As Jefferson (2008, p. 370) observes, we need psychosocial approaches because "some sociological questions require psychological answers" and "the psyche dynamically alters that which society imposes upon it". These observations seem particularly pertinent in exposing what is missing from these very influential contemporary social theories.

Within fandom studies, enacted affinity involves pleasurable repeated engagement with the object of the fan's attachment across different mediums. This idea continues outside of fan studies. For instance, film theorists argued that one of the reasons the spectator returns to the cinema was to repeat the pleasurable cinematic experience (Bellour, 1974, 1988; Mulvey, 1975/1988). Generally then, returning to a text or repeated viewings have been understood in relation to pleasures and possible fascination of the text, and the ability of the cinematic apparatus to assuage primitive forms of anxiety (Mulvey, 1988; Stacey, 1988). As I have already explained, however, viewers often return to texts that they do not find pleasurable (see Chapters 3 and 4). Psychoanalysis is not a unified body of theory and there are different schools of thought, but most would agree that the idea of pleasure is more complex than mere enjoyment. Pleasure and pain are often linked for example. Lacan's (2007) notion of jouissance is, on one hand, associated with the drive to seek pure bliss and ecstasy, unmediated by the demands to temper this in the face of the consequences of reality. Equally, it is linked to the idea that it is possible to go "beyond the pleasure principle" in this quest, only to find that pleasure is painful and disappointing. Prior to Lacan, Freud (1920, p. 18) noted that life experiences such as trauma can cause an individual to repeat behaviours which do not satisfy to avoid remembering something difficult that belongs to the past. I will return to this in Chapter 4. Some viewers acknowledge that a visit to the cinema is a superior sensory experience and hence pleasurable (Sue, Chapter 5). Others dislike the behavioural protocols associated with cinema and prefer watching films at home (see below Daniel, Chapter 3). Therefore, the idea of cinematic pleasure only partially explains a viewer's reason for returning to a text. The idea that the text is fascinating is, I think, more interesting, but the nature of this fascination needs exploring beyond accounting for the visual and formal workings of the text.

Some work has been done on the use of technology and its potential to offer opportunities for repeated pleasurable viewing, in particular the use of the VCR (Woods, 1993). The use of the re-wind facility was crucial to Walkerdine's (1986) analysis in *Video Replay* as the working-class family man she encountered in her research kept re-winding the video to watch a violent scene over and over. Walkerdine linked this in her analysis to the role of unconscious fantasy in this man's struggle against the system and to conflicts in his masculine identity as a working-class man. In this study we see the notion of pleasure and textual fascination given more depth through the use of psychoanalysis as it points to the use of the VCR to address psychic and social conflicts.

In psychoanalysis, repetition, as suggested above, is associated with trauma and loss. Bainbridge and Yates (2007) examined this in their study of the use of video/DVD and repeated viewings. They found that pleasure and trauma are often closely related and, at times during repeated viewings, audiences experience this notion of something not-pleasurable ("It makes me feel bad") that I have described previously. This study focuses on the relationship between repeated viewings of films, which foreground masculinity and the male viewers' use of DVDs which offer highly sophisticated viewing options. This development is located within marketing trends in the reissuing of classic movies, in forms that allow male viewers to take control of the text, in ways that might appease the conflicts of the contemporary crisis of masculinity. In particular, this promotional trend is said to create nostalgia for traditional forms of masculinity that men may fantasise have been lost. The study is valuable because it shows that identity, pleasure and trauma can be very closely linked in the viewing experience, and this has both personal and wider cultural resonance for viewers (see Chapters 3 and 5).

Stacey (1994) has argued it is not possible to capture the experience of the original moment of viewing. All that can be obtained are "retrospective representations" and interpretations of the moment and initial responses. However, the original moment may not be the *most* significant. As I will show, texts can take on new meanings in the process of revisiting and repeating viewing. Seemingly benign texts can acquire unexpected emotional force in subsequent viewings which psychoanalysis can help explain. Sutton (2004), for instance, proposes all spectators carry a personal cinema history which comes into play when audiences read films. I would broaden this to include textual history including television, literature and art. This history is constantly returned to and re-translated, in the light of biographical experiences and events. Sutton (2004) adopts the concept "apres-coup" also referred to as "afterwardsness" from the psychoanalytic theorist Jean Laplanche[3]. This concept is commonly used in theories of trauma but can also be used to understand memory and temporality which are pivotal in relation to the complex processes associated with repeated viewing. According to Laplanche trauma has two moments, the first being the actual event, which is not perceived as traumatic at the time, such as a child being confronted by the primal scene, where the child becomes aware of his parents' sexuality. Initially the child does not interpret the scene as traumatic because he lacks the developmental understanding to realise fully the sexual nature or consequence of the scene. The second moment comes at a *later*, more developed stage when the child realises that what he saw was traumatic. The important idea here is that the past is always interpreted from the current position the individual is in. Experience and the memory are changed in the light of current experience and knowledge. However, this is a kind of to and fro movement where the past is in the present but the present shapes the memory of the past. This can be used to explain the shifting identifications I described in the introduction. This is also why there cannot be one definitive reading of a film; our readings of films and television texts do not stay fixed but move and change over time whilst retaining some core features. In fact, afterwardsness is a useful way to think about responses and meaning in different viewings of the same text. The viewer is always in a different position, changed by the present and other experiences between viewings. Other forms of evidence coming from the child's/adult's experience can lead to a past event interpreted in the light of the present, as anxiety-provoking and traumatic. In Chapter 5, for instance, I will return to the example given above of Mary's response to *Father of the Bride*. I will show that Mary's strong emotional response involved a long history of her own intense relationship with her father that this textual moment captured. This was reactivated in the research interviews and also in the light of the impending marriage of her idolised only son at the time the interviews took place.

Another relevant concept is repetition compulsion[4] which refers to repetitive behaviours or memories which repeatedly go over a particular scenario and prevent mourning for someone/something significant that has been lost (Freud, 1899, 1920). Screen memories, for instance, are an example of the way some traumas are dealt with defensively. Here, a seemingly innocuous memory (perhaps from childhood) is associated with a strong emotional response that is not fully understood by the person, nor is why the memory should merit this response (Freud, 1899, 1901; Rosenbaum, 1998) (see Chapter 5). For Lynne Layton, traumas associated with lived experience often take the form of repetition compulsions which are also "lived" as "character" (Bollas, 1992; Layton, 2004). Our habitus, the things we like or dislike, can be manifestations of this, as we see in Chapter 4. This can help explain why Sophie watches films that make her feel "bad" or why Chris in Chapter 3 engages in viewing which he finds threatening. These examples may not point to specific traumas, (although there is evidence of trauma in some cases), but exist as examples of what Bollas calls the "private use of the object" (1987, p. 26). In this personalised object use, character can also be seen as the manner by which individuals manage personal emotional, social difficulties and loss. These viewers utilised objects and practices in moments of emotional struggle. Layton (op. cit.) proposes that the return to a repetition scenario is not just the wish to go over the same trauma again and again, but is motivated by the wish to find a way out of the problem and to promote change, however minute. Later chapters will offer in-depth exploration of instances of favourite or significant textual moments in impact texts, as repetition scenarios. I will argue that they feature in attempts to work over past and present difficulties that individuals face. This also suggests some of the depth that is missing from existing models of needs, uses and audience responses.

Thus, record/replay/time shift technology can allow us to explore the way texts are used in relation to the way the self is formed, reformed and worked on, through films and television programmes. Clearly, the arrival of this technology has added many new and complex layers to spectatorship. At this point, I want to develop these ideas further by introducing perspectives from object relations that can help explain some of the emotional processes involved in object use.

Texts, viewing and evocative object use

Bourdieu's work was important in that it noted the social significance of people's use of culture and cultural objects. In this section I want to develop this and pick up on the ideas introduced in the opening chapter, to explore why relationships with objects and things such as texts, practices and favourites are "important" subjectively (see Winnicott, 1971; Bollas, 1988, 1992; Phillips, 2002). I will explore this through a discussion of the work of Donald Winnicott and Christopher Bollas, who has become one of the most notable contemporary theorists to develop Winnicott's ideas. Also I shall discuss Roger Silverstone's (1994) original and innovative adoption of Winnicott's work to understand the role of television in everyday life. I will link these ideas to the questions raised by object relations theories for theories of identification in viewing. I will begin with the idea of the transitional object.

Object relations propose that humans are object-seeking and this pursuit starts in infancy and continues throughout life. Human objects are complex, sensuous, affective beings, and when object relations talk of successful object-relating as a feature of emotional well-being, it is interested primarily in the quality of the relationships with other human objects. Therefore, the use of texts and cultural objects will be discussed in later chapters in the light of this recognition. Winnicott (1990, 1971) asserted that the conditions of the maternal environment in early life laid the foundations for later object relating (see also Phillips, 2002). In particular, he was interested in the stage of development in infancy where the child has to learn to separate from its primary caregiver (usually the mother) to become a confident, emotionally secure individual. The main developmental achievement for the child involves moving from a state of absolute dependence on its mother, whom it initially views subjectively as an extension of itself and under its omnipotent control, to one where the mother is objectively perceived as an independent being. It is what Winnicott (1971) called "Transitional Phenomena" and "Transitional Objects" that make this move possible. The process takes place in the "Transitional Space" (Winnicott, 1971) where objects, including material objects, play a vitally important reality-testing function. Winnicott (1971, p. 3) says: "This is a space that is neither inner nor outer, but a third or intermediate area of experiencing to which inner reality and external life both contribute."[5] Thus this space is neither subjective nor objective and it provides both continuity and a bridge between inner and outer worlds. In psychosocial theory this space has been called the hyphen that some theorists adopt (psycho-social) (Hoggett, 2008). I will discuss the implications of this for research in the next chapter.

The mother, on whom the child is dependent, is responsible for managing the infant's environment and meeting all its needs, and Winnicott stresses three essential "functions" of the mother in relation to this: "holding, handling and object-presenting" (1988, p. 38). Prior to the transitional stage of development the baby treats the objects it encounters as extensions of itself, in a form of "subjective omnipotence" (Winnicott, 1971, p. 95). In the case of object presentation, when the baby wants feeding for instance, it is his mother who senses this and responds to his needs, presenting him with the breast to appease his hunger. The infant, however, phantasies it has imagined this satisfying object and then "as if by magic" it appears (Phillips, 2007, p. 114). However, this is not realistic, and the child must come to recognise its mother as an object not under its magical control. It is the use of objects in the transitional space that helps facilitate the change of relationship with objects. Winnicott (1965c, p. 45) summarises that "here the infant changes from a relationship to a subjectively conceived object to a relationship with an object objectively perceived". As part of this process the infant begins to "weave other than me objects" into its experience (Winnicott, 1951, 1971).

Instead of meeting the child's needs immediately the mother begins to gently frustrate the infant, for instance by forcing periods of separation upon the infant, which has the result of causing frustration and fear associated with the realisation he is dependent upon another. These fears are managed through the use of transitional phenomena. The child selects a toy, a teddy bear, or a doll for instance, or a blanket, which becomes the first "not-me" object (Winnicott, 1951, 1960/1965). The object comes to symbolise the presence of the mother, and its presence soothes the child in her absence until she returns. The mother colludes in the child's fantasy about the indispensable nature of the object and in the child's illusion that it magically created the object. Although it is likely to have been presented to the child, this object cannot be imposed upon the child, nor can it be given to him as a transitional object, "but only as something he may choose to turn into one" (Phillips, 2007, p. 115). It is the mother's pivotal role in object presenting, however, that led Adam Phillips (2002, p. 54) to summarise that we start with the mother then later when we enter culture we use objects to unwittingly "bring parts of ourselves to life". Thus the Transitional Object is always a combination "of subjectivity and objectivity, but one that provides by virtue of being more than the sum of its parts, a new third alternative" (Phillips, 2007, p. 114), part of this intermediate, or third, space.

In the transitional space, the child uses objects creatively and through *play* experiments with moving between the subjective perception of objects as an extension of itself and objects "objectively perceived" (Winnicott, 1971, p. 4). In the fantasy of play the child tests out its own omnipotence, power and vulnerability. The child can use the object, symbolically, to test out such powers as it stands for the mother. It can throw the object away, hit it, project hurtful impulses and wishes at the object and, like real mothers, it survives. The silk at the end of the blanket always feels the same, it retains its qualities and, like the dependable mother, returns. This helps to build the child's confidence. At the same time, however, there is the realisation that individuals are always vulnerable, to the extent that their survival is ultimately linked to other social beings. This is another reason why human objects have much more crucial significance than cultural and inanimate objects however much we might invest in them (see Chapter 4).

It is these ideas of creativity and play that have made Winnicott's work attractive to some cultural researchers (see Hills, 2002). Transitional object and transitional phenomena usage are present throughout life and can be utilised in all situations where creativity of thought and perception are needed or to manage anxiety. The fundamental insight of psychoanalysis is that: "the infant remains within us, the unconscious is timeless and must be catered for – it does not disappear with our access to maturity...the infant plays an essential role, in a modified form in our adult life" (Craib, 1998, p. 53). Hence the infant and the unconscious traces of its experiences are present when we engage in viewing.

Moving specifically to television, Silverstone (1994), evoking transitional processes, stresses the importance of communication for individuals who are both product and producer of symbolic acts of communication. He explores television as an everyday activity in the light of the changes associated with post-industrialisation, neo-liberalism and in relation to the need for what Giddens terms, "ontological security" (1991). Silverstone (1994) uses Winnicott's ideas to show how television can provide forms of emotional and ontological security. This involves two elements: one, the idea of TV as physical object, the box, that is part of everyday life: and two, the constancy and dependency provided by the "flow" (Williams, 1974) of broadcasting. Silverstone (1994) states that television, as a pre-packaged complex communication of sound and image, makes powerful claims on reality and emotion. He argues that television operates as a transitional object and can evoke the experience of the mother and the management of the maternal environment in infancy. He lists the following transitional characteristics:

- Television is always there, constantly available.
- It can be used as a babysitter.
- The perpetuity of sound, voices, music and flickering images is soothing.
- It provides continuity through the constant flow of programming. There is a reassurance obtained through the predictability of the weather, news, or soap operas. Threats associated with the real world as well as emotional anxiety will be addressed and also managed on behalf of the viewers. Television is an object of trust where reality testing can take place in the safety of the home (see also Scannell, 2007).
- Television, like the teddy bears and blankets, can survive viewers switching off in anger, shouting abuse at the set, even hurling objects at the set and, when switched on again, it remains the same (Silverstone, 1994).

Watching television in this model is seen to be a wish to return to the comfort of early fusion with the mother, which is very similar to the regression to dream-like states of 1970s film theory.[6] Whilst this accounts for the transitional function of television per se, it does not account for the processes that are involved in specific texts becoming special and significant. In later chapters, therefore, I will explore the idea of specific texts and viewing practices as Transitional Phenomena (see Chapters 3 and 5). I will also develop the idea that the transitional space is an area where reality testing takes place and examine the imaginary and emotional work that might take place when viewing favourites.

Christopher Bollas (1987, 1988, 1992, 1995, 2002) is interested in the role played by object use in allowing character to be articulated. I suggest that rather than think of favourites in relation to questions of taste, they are in fact expressions of character which also accounts for internal processes associated with object selection. Bollas develops the idea of "True Self", a category Winnicott (1965a) devised to capture the "essence" of life. This should manifest as a sense of aliveness associated with a sense of feeling real, expressions of spontaneity and a sense of contentment, where the individual is happy to be alone with himself/herself. It covers what is distinctive and unique to each person and hence is articulated as character (Bollas, 1995).

This ability to live spontaneously and use objects creatively in response to desires, needs and imagination is, in adult life, the sign of emotional health and results from "good enough" management of the maternal environment that I described above (Winnicott, 1965a). Bollas (1992) argues that people invest objects and practices, such as viewing or listening to music, with unconscious meaning. When the object or practice is put to use individuals' psychic history is evoked. For Bollas (2000), psychoanalysis is a "theory of memory's desire, of experiences that having yielded a certain value become the basis of subsequent related interests" (Bollas, 2009, p. 18). Some experiences enhance self-experience and, in some cases, might be pleasurable, but in others can involve pain. Thus, certain cultural experiences can be returned to and utilised by the individual in the process of developing unconscious structures which Bollas calls "psychic genera". Such structures allow subjects to experience aliveness and psychic growth, in keeping with the spontaneity and individual uniqueness of true self living, and contribute to psychosocial meaning making.

Bollas uses the term the "unthought known" to refer to forms of unconscious affective and embodied knowledge which defies linguistic articulation and that informs object selection (Bollas, 1992, 2000). At certain moments we experience embodied reflection or heightened self-awareness. This can lead us in the direction of particular objects and/or experiences, (perhaps explaining the move towards doing something un-pleasurable). For Bollas, psychic life and survival revolves around the desire to know, explore and revisit pivotal moments from life with the aim of having some form of self-transformation, however minute.

In the encounter with the object, the self-experience which Bollas (op. cit.) calls "idiom" is dissipated. Idiom involves a mixture of the earliest experiences in infancy, life experiences and events, and our inherited disposition. There is a forwards/backwards motion associated with idiom, where individuals move from a recent to a distant past and back to the present in response to a particular stimulus. This stimulus, in relation to viewing, as I will discuss, involves particular forms of identification. For Bollas (1992) once an object such as a text, or an activity such as watching television, has been selected for use, the unconscious begins the work of employing the object subjectively. Mann (2002, p. 58) explains: "these various objects of usages are not simply a static or dynamic representation; each object represents a different link by which the subject employs the object and in doing so the subject defines and extends itself." The psychic structures generated through creative living promote the ability to withstand trauma and assist in reparation throughout life.

Bollas (1987, 1989) explains that objects hold out the potential to alter and change us, because our encounters bear the traces of our first object relation with our mother/primary carer who is a "transformational object" because she transforms our environment. Mann (2002, p. 60) argues that throughout life individuals long "for the ability to submit to objects that will change the subject's self-experience". One of the ways this happens with objects is through processes of projection, as the mother functions like a container into which the infant projects bits of itself. Similarly O'Shaughnessy, proposing a theory of emotional responses to films, argues that we "project our character onto the things around us" (1994, p. 45). He contends "our favourite films, like our dreams, offer a reflection of our personal and psychic concerns", and our reading of certain films is not entirely a reflection of the film text but of "the self" and our memories and biographical experiences. Objects can also be used as containers for different emotional self-states, and in adult life we use objects as part of a *cumulative* process of self-elaboration, returning to objects and experiences because they allow us to elaborate psychic idiom and they help us to perform emotional work on memories and experiences.

Grotstein (2002) argues that whenever we encounter "objects" there is always an unconscious need to personalise them and endow them with our own subjectivity. However, objects also "stimulate" us (Bollas, 1992, 2000), which I suggest can be thought of as a kind of hailing or interpellation. These *Evocative Objects*, like psychic "keys", open doors to unconsciously intense and rich experience in which we articulate the self (Bollas, 1992, p. 17). Evocative objects often share a characteristic of favourites in that they involve an element of surprise when the object first hails the subject, as seen in a number of the examples given previously. Bollas (op. cit.) calls this entity an *Aleatory object* (see also Hills, 2005).

Conceiving of favourites as evocative and aleatory objects, it is useful to also look at other features of the ways in which objects stimulate us to use them:

- Sensationally (materially, affectively and embodied).
- Structurally (*use* potential promotes experience on the basis of the character of the object and its *use value,* thus we can think about the formal nature of cinema and television as mediums and the construction of specific texts).
- Conceptually (concerned with thinking and thought both conscious and unconscious, and involved in meaning-making processes).

- Symbolically (as a container, selectively endowed with past experiences and self-states through projective identification and I would argue this links particularly with the following type).
- Mnemonically (concerned with memory generation and where a present object or an object from the past stores a self-state from a particular time).
- Projectively (mnemonic objects may be evocative of the past but are used projectively in the present to form association which allow the self to project itself in the here and now) (see Bollas, 1992, p. 34).

Certainly films and television have complex *use* value in that they work predominantly in the area of the first three features above, in particular affective, embodied, meaning-making processes that are designed to move the viewer. It is, however, the inclusion and intensity of use in the latter categories which accounts for their designation as an evocative special memory or impact text, as later case-study chapters will show. Thinking more about these categories, with specific reference to films and television viewing, I want to add that the question of the intensity of multiple forms of object stimulation can be understood in relation to two forms of identification, in particular idiomatic and biographical identification or a combination of the two.

It is useful to adopt an idea from Laura Mulvey's (1975) notable work on *Visual Pleasure and Narrative Cinema*, concerned with the way the narrative momentarily halts to allow the camera to focus on the female form, thus giving the viewer visual pleasure. In the aleatory moment, when a viewer encounters a text that has a significant emotional impact, the *viewer* momentarily halts the narrative. At this moment they temporarily leave the narrative, which will move on, but the viewer will not move forward with the narrative. Like Mary above, they will be caught in a moment of emotion or they find that a strong memory shifts their thoughts away from the text. Here its use value changes from one of entertainment, as a multi-layered set of processes come alive, including conceptual, projective, mnemomic and symbolical work, turning the impact text into an evocative object. So whilst *Father of the Bride's* use value, which Mary referred to as "harmless" might suggest sentimentality, laughter and fun, the text's use value was altered once it became an evocative impact object. In this case, it becomes a melancholic object which involves projections of past and present difficulties, fears and loss, and this is explored further in Chapter 5. At the same time, it can capture an embodied feeling of aliveness, perhaps because it evokes a memory or sensation from early life, which might contrast with the emotion of the text, but makes the text an object of pleasure, and thus a favourite, because of this combination of idiomatic and biographical identification and heightened self-awareness. Evocative objects may or may not incorporate transitional characteristics (in that they can be soothing and allay anxiety). The point is that each case of object use can be assessed individually. As I will explore in Chapter 3, it is the specific combination of text, context and viewing practice or style that designates the use of a text as a transitional object. The aim of my discussion here is to explain why pleasure is a misleading and rather one dimensional concept, as object use is complex and the experience of returning to particular texts may be intense but it is not always *pleasurable* in the conventional way pleasure is understood.

I also want to briefly address identification, which in conventional psychoanalytic use points to an emotional connection with an object. Freud noted that identification is based on some form of recognition or perceived similarity between the individual ego and the object (Freud, 1921; Hinshelwood, 1989). The idea that audiences identify with texts has been an important concept in audience studies concerned with understanding the ways audiences make meaning in relation to the texts they view. If we think of idiomatic identification in response to object stimulation, we necessarily address the idea that particular text selections and viewing styles are examples of the ways in which the self speaks. They carry a history and bear the marks of parental provision and internal object relations (Bollas, op. cit.).

Identification is not just between the self and external objects; it can be a psychic process associated with internal objects. Melanie Klein (1940, 1946) describes the inner world as peopled by representations of good and bad objects and part objects which are the result of primitive emotional processes such as projection, splitting and unconscious phantasy. The term "internal object" means a "mental and *emotional* image of an external object that has been taken inside the self", and the "character of the internal object is coloured by aspects of the self that have been projected into it" (Melanie Klein Trust, 2014). These mental representations are cognitive affective units that are composed of a self-representation, an object representation and a "linking *affect*" (my emphasis) (Kernberg, 1976, p. 163). The representations are not facsimiles of real people living in our inner world, they are traces of our *experiences* of them. Thus significant relationships and encounters with objects of earliest life leave their traces as "psychic textures" which Christopher Bollas (1992, p. 59) says are "ghosts of others who have affected me" and "we are internally transformed by objects that leave their traces within us" (people and films/programmes/visual images, etc.). Our inner world is peopled by the psychic and phantasised traces of family and other key figures whom we also relate to, but in two ways: as self and other. That is what it felt like to be me in relation to the person or object and other what I imagined (in phantasy) the other felt and imagined in relation to me (Redman and Whitehouse-Hart, 2008). We are not conscious of this and often only aware of such processes as emotional textures or mood. I think this puts a rather interesting twist on an important idea in television studies: family viewing. This suggests that families are present in our psychic textures when audiences are viewing without needing to be physically present.

Significantly Bollas argues that "the decision to use an object rests with the unconscious aims of the person" (1992, p. 36). I would caution that although Bollas (ibid.) is aware that all activities operate in different "hermeneutic spaces", pointing to the way activities carry social meanings, he prioritises unconscious motivation at the expense of understanding that the social and economic domains partially *define* and also *limit* the possibilities for object selection. This shows the limitations of his work to fully meet psychosocial aims.

The home as a psychosocial space and TV as a psychosocial object

Television is often designated as a passive, low intensity medium, where viewing is distracted, watched with a glance rather than a gaze (Ellis, 1989). Viewing is described as wallpaper, as something insignificant. In contrast, this section explores watching television at home as a powerful intimate, affective and emotional experience.

Bollas argues: "We can learn much about any person's self-experiencing by observing his selection of objects" (1992, p. 32) and that the objects a person selects are a "valued part of the fieldwork" (1989, p. 18). The domestic setting or the place/places we refer to as home is our most personalised space. It is an environment where objects are gathered together and selected for use as vehicles through which the subject articulates himself/herself. In a recent anthropological account, Daniel Miller (2008) explores how people express themselves through their possessions and their homes, and how those possessions play a role in our relationships with self and others. He concurs that objects are a kind of language which speaks or paints the self for the outer world. For both Bollas and Miller (2008) objects are important methodological tools in learning how to "read" people through their possessions. Miller also shows how a discussion of "things" in the home inevitably conjures up past biographical experiences and memories. The case study chapters will explore this further.

Viewers enjoy going to the cinema but often prefer viewing at home. Daniel explains in Chapter 3, it is possible to "turn a film off". There is something about the home setting that means there is more possibility for viewers to be themselves. Sue (Chapter 5) says, "It's funny because I say I prefer to see things at the cinema yet the ones that have had the biggest impression on me I've watched at home." It is those films that have left an impression and that she calls favourites.

The home as a setting carries with it the history of previous homes in the individual's life and the internal object relations formed in relation to the people, usually family, and the experiences that took place there. Viewers, like those who appear in later chapters, who were born prior to the digital age, also have the memory of living in a house with only one television set and three channels to choose programmes from, which was always a process of negotiation. Viewing was predominantly a quotidian shared experience and also an event, as extended family and community came together to watch sports and historical events. Mary recalls watching the FA Cup Final with her extended family packed into one room. Daniel recalls Christmas Day viewing *Noel Edmonds Christmas Day Special* (BBC) with the family.

Television acts as a backdrop to the rhythm of the activities of daily living, which also form part of the experience. Sometimes viewers attempt to *recreate* features of these patterns in adult viewing styles (see Chapters 3, 4 and 5). Home represents many things; it is a place where we can relax – a place of safety, joy and intimacy – and at the same time it may be an intense and oppressive place. In childhood, home is the canvas where identity is painted through discourse and processes associated with the environment as discussed, and in relation to key figures such as parents as detailed in Freud's Oedipus complex.

Television viewing in childhood is often associated with intimacy. Sophie (Chapter 4) has happy memories of having a bath on Sunday evenings and the family watching *The Muppet Show* together. Chris recalls his enduring love of slapstick comedy starting with a season of Jerry Lewis films on television, watched after school, which various family members would dip in and out of. He liked the films, but the pleasure, which included enjoyment in the conventional sense, was also that they were viewed against the rhythm of the house:

> We wouldn't always sit down and watch them together but my family would always be about.... if I was sitting watching, my Dad would probably just be coming home from work and he'd come in when

> I was watching... And my Mum'd be buzzin about doin' something. (Chris)

In a similar vein Lacey (1997, 1999) recalls watching Hollywood musicals at home with the sounds of her family talking, and the smell of lunch cooking. Ellen Seiter remembers watching soap operas during her mother's serious illness and how this gave her stability (in Brunsdon, 2000). The emotional highs and lows of a text can chime with the mood of the home and the life events taking place within it. It has been argued that television can operate as part of a process of relay, through which difficult feelings and emotions within families can be articulated discursively through the TV and the text (Walkerdine, 1997). This is part of the emotional sensibility of television viewing at home.

Recently television has been theorised more systematically as an *affective* genre. Most theorists make distinctions between emotion and affect. In social science, emotion is a term often used as an equivalent of feelings. Emotion refers to conscious experience and is something that becomes real as it is named linguistically, such as anger, shame, joy, etc. Affect is connected with something more basic, associated with the perception of bodily states, and unlike emotions, these states have not been "mapped by the brain" to become "feelings", which can then be worked consciously into emotion (Ducey, 2009, p. 191). Affect can also be associated with mood and often defies language (Day Sclater et al., 2009; Ducey, 2007; Gorton, 2009; Kavka, 2008; Spezzano, 1993). Despite debates about the level of distinction, emotions and affect represent a "bridge between the individual and the social", and we can think of both terms as "quintessentially psychosocial phenomena" (Day Sclater, 2009, p. 1).

Thinking about this idea, of television as a backdrop, Misha Kavka's (2008) work on the affective valence of television is useful. For Kavka (2008, pp. 33–34) affect has "object potential", and it is "mobile" and transmittable, moving between people and things and having the ability to flood a room. Affect has a "point of emergence" on what has been called a "cusp formation" (Massumi cited in Kavka, 2008, p. 31) for the particular material forms the feeling may inhabit, television for instance. Television is also an *intimate* medium with the ability to "bring things near". Thus, if we combine the use of close-ups, television's ability to present scenarios for viewer's to identify with, and the displays of emotion seen on the television screen with the intimacy of the home setting and all that takes place therein, we find, then, that television in the home creates a space for the viewer to explore "things that matter" (ibid., pp. 36–37). Often this takes place through forms of affective identification with what is on the screen and viewers become bound by an "affective glue" (ibid., p. 37).

Kafka's insightful theory was designed to account specifically for affective responses to the reality television genre, well known for its depiction of explosive emotion. However, the point I want to take from this is the idea that the intimacy of the medium, moving affect and objects within the home, combine to create the conditions for what I would call a psychosocial cusp formation. This makes television a particularly affective and emotionally productive medium and object. To this, I would add the psychoanalytic idea of transference, arguing that our inner world is also generative of affective responses. Our contemporary psychic reality and unconscious idiomatic aims to personalise the world accompany the viewer to the television screen as part of this cusp formation. It is this inner psychic reality that carries our history, internal object relations and phantasies and which dictates the nature of the affective experience, how affect moves into feelings, and they, in turn, become experienced and named as emotions.

This chapter has outlined the characteristics of favourites, pointing to their biographical significance. Psychoanalytic theory, particularly insights from object relations, was introduced to help explain viewers' attachments to texts which evoke heightened self-awareness with all its biographical and affective significance. The emotional role played by repetitive viewing that replay/time-shift home entertainment technologies facilitate was also discussed. This led to the conclusion that the home is an intensely affective psychosocial space and within this space television is a quintessential and evocative psychosocial object. The next chapter will develop this through an exploration of research methods for psychosocial and audience research.

2
Psychosocial Methods and Audience Research

The previous chapter argued that the experience of watching favourite films and television, particularly in the domestic or home environment, should be understood psychosocially. Within Communication and Media Studies, prior to recent developments on viewing and affect, the idea of researching unconscious communication was eschewed. Established approaches in empirical audience research have generally started from the assumption that it is mainly cultural and social differences, such as class and gender, that are significant in knowledge production (Seiter, 1990). Making audience research on the interpretation of meaning and identity largely focussed on processes that work one way: outer to inner. As a corrective to this, Chapter 1introduced ideas from psychoanalysis to offer some perspectives on inner processes, such as unconscious personalising of texts and the idea of an intermediate area of experiencing. Reality is not something objectively perceived, but something that is also subjective and with internal dynamics. Inevitably, this has implications for epistemologies that "draw a line between subject and object" and the social and the psychic (Young, 2009). Psychosocial insights inexorably point to the need to review established methods and interpretive paradigms and epistemological assumptions. This chapter offers, therefore, an account of a psychoanalytically informed method used in a study which represents one of the first attempts to apply such new approaches to audience research. The study and method are distinctive in that they overcome some of the divisions between text-/audience-based research and social versus psychoanalytic (textual) interpretations.

This chapter describes the epistemological assumptions underpinning the research design and introduces relevant theoretical concepts, ideas and the method on which the research was based: Hollway and Jefferson's (2000/2010) Free Association Narrative Interview (FANI) method.[1] The following section of the chapter re-evaluates a well-known piece of sociologically influenced audience research, to demonstrate that psychoanalysis can contribute towards the production of psychosocial knowledge that enhances our understanding of audiences, texts and genres. The purpose of this is also to apply psychoanalytic theories that are concerned with the production of knowledge to established audience research.

From the outset it is important to note that concern has been expressed about the viability and ethics of adopting a set of ideas and techniques designed for a specific use, the treatment of emotional disturbance in a clinical setting, and applying this to the arena of academic research (see Frosh, 2010; Frosh and Baraitser, 2008). At this point I will signal to the reader my recognition of this concern and note that I adopted established procedures for the protection of research subjects including obtaining written consent, anonymising and explaining that interviewees had a right to withdraw, stop the interviews or refuse to answer questions at any time. However, Chapter 6 deals with the ethical issues in context through a case-study; therefore, they will not be dealt with here.

Psychosocial research subjects and knowledge

In accounts of identity in audience studies, frequently the human subject is the end product of outer processes, socially constructed through discursive structures, disciplinary practices and unequal power relations. Thus, what constitutes an "individual" is socially and historically specific (Henriques et al., 1984; Mansfield, 2000). Psychoanalysis, as a corrective to this, implies that whilst human beings might appear to behave and think rationally, reality is never experienced fully objectively, as unconscious processes mediate our thoughts and influence our actions. Despite this, even film theory's enthusiastic take up of psychoanalysis gave weight to the constitutive capabilities of representations and "film language", ultimately positing a subject made in relation to language and thus a "social subject" (Craib, 1998, p. 10). Some theories might appear to adopt the language of psychoanalysis and pay attention to the inner world. For example, subjectivity has been said to be shaped by disciplinary practices inscribed on the soul (Rose, 1999) and contemporary society and culture motivated by narcissistic needs for recognition (Lasch, 1979; Biressi and Nunn, 2005; Ouellette and Hay, 2008). On closer inspection, however, the emphasis remains one-way, socially determinist, concerned with social forces and their effects on the inner world. The way in which the social world is taken up and affected by internal processes is not fully explained in these accounts.

Thus, a psychosocial approach to the subject acknowledges internal processes. So, for example, in a Kleinian account, the inner world, made up of internal objects that are representations of things; people, families, experiences and ideas, is constantly in flux, with objects related to as good and bad[2] and subject to processes of phantasy. These are not purely intra-psychic processes, however, as the events of the outer world are also perceived through the lens of phantasy which is present in all conscious and unconscious mental activity and perception (Mitchell, 1986; Isaacs, 1952). These outer events change the dynamic relationships of the inner world, through processes of projection and introjection. Consequently, it is never possible to experience others (and animated objects such as films and television) in research, or any other situations, without the presence of varying degrees of unconscious phantasy.

It is useful to think about the characteristics of the internal world, such as the idea that families exist in the inner world and are related to, to understand its role in identification and the creation of meaning, which are central ideas in the case study chapters that follow. The experiences of dependency in early life are frightening, threatening and frustrating for the infant, and also leave their traces on the inner world. As the reader will see later, these experiences, marked as they are by anxiety, frustration and aggressive responses, occur as the outer world refuses to conform, and this places limits on the child's omnipotent desires and wish to control. Anxiety is, therefore, another pervasive feature of self-experience (Klein, 1929, 1946, 1948). Both anxiety and omnipotence are features of self-experience throughout life. I will now discuss how research creates a fertile terrain for particular types of anxiety and omnipotent thought.

Defence mechanisms are employed to protect against anxiety which is generated in response to threatening experiences. Splitting is used to preserve the good by keeping it separate from the bad, for example (Klein, 1946; Segal, 1992). When anxious, the infant may also project or place something (usually an unpleasant or bad feeling) into another object. Introjection involves taking-in parts of something good and making it a part of the self. Later chapters will think about projection and introjection in relation to film texts and the television object. It is the child's need for omnipotent control that motivates the use of unconscious defence mechanisms. This is one of the ways that difficult and overwhelming feelings, of dread for instance (Klein in Mitchell, 1986, p. 133), are managed. If a child can master and control his objects, then the object cannot threaten the child. These "primitive" (Hinshelwood, 1989) defence mechanisms are associated with the paranoid-schizoid position, which is characterised by anxiety and phantasies of persecution and aggression. The child is not concerned with any notion of reality, which can be objectively perceived. It is only concerned with its desire to appease its anxiety and allay fears of annihilation, hence the use of defence mechanisms. As the child develops, it learns to tolerate good and bad in objects without feeling that safety is compromised and with the need to be in omnipotent control. Klein calls this the "depressive position". Here, persecutory anxiety is replaced by feelings of guilt for the aggression and damage done to loved objects in phantasy (Segal, 1992, p. 38). This represents a more mature position where efforts are made to "maximise the loving aspect" of the relationship with "whole", rather than part objects (Ogden, 1989). Most notably the infant is able to recognise its mother or primary carer as a separate object with her own wants and needs. Thus, there is a shift from a relationship with "objects" in the paranoid-schizoid position as the relationship develops to one between "subjects" (Ogden, 1989). This can also be associated with the transitional space introduced in Chapter 1. Here the child oscillates between subjective omnipotence, where objects are an extension of his/her self, and seeing objects, principally its mother, as a separate object, with her own character and needs, "objectively perceived" (Winnicott, 1971, p. 4).

Another development is that the possibility of reparation and the ability to tolerate ambivalent feelings means that reality no longer needs to be brought into line with an anxiety-dominated inner world, but can be tolerated and therefore perceived more realistically. Ogden (1989) notes that the paranoid-schizoid and depressive positions are not developmental stages; we *oscillate* throughout life between both positions as responses to experiences and change. Following this, below and in Chapter 6, I will show that an understanding of omnipotence, as a result of anxiety generated for researchers, can result in researchers unconsciously trying to control object interviewees and that interviews can generate some of these primitive anxieties that I have described.

The important point to be gleaned here is that learning (knowledge acquisition) is an affective, as well as a cognitive, process. This is significant for audience research which is interested in the question of meaning but has often been focussed on meaning and how it is constructed in texts. Theories of decoding have not considered how affect (which might also be elusive and difficult to capture and articulate in language) might inform the meaning that that viewer gives to the text that they are watching. Similarly, it is important to recognise that a psychosocial understanding of meaning is one that is produced out of the flux and tensions between inner and outer worlds. It has the qualities of the transitional space in that it "always comes from both without and within and is an inextricable mixture of socio-cultural and historically contextualised on the one hand and the personally psychodynamic and biographically contextualised on the other" (Chodorow, 1999, p. 3). This is a radical view of reality and the world, which is elusive to traditional forms of positivist scientific observation based on cause-effect logic.

Thus, post-Kleinian object relations and relational psychoanalysis propose that ordinary everyday experience is conflictual, anxiety-laden and that "relationality", "specifically unconscious intersubjective dynamics", are the "founding principles" for identity transformation and transition (Hollway, 2009, p. 217). The most significant point for audience research that I want to take from this, is that these unconscious conflictual dynamics are significant in that "they do not observe the boundaries of the physical individual but move between and among people without their intention or awareness" (ibid., p. 218). We are in others and others are in us. This is also linked to a "life-historical time perspective" pointing to identity as something not fixed, but as something that, throughout life, is always in the process of "becoming" (see Chapters 3–6). It is also worth noting that the emphasis placed on the relationships with objects extends Freudian theories which were based on drives and where the object (person or thing) was a "target" selected for the satisfaction of desire or pleasure deriving from an instinctual impulse (Freud, 1920).

The idea of unconscious intersubjective dynamics has implications for research, particularly for objectivity, which is the cornerstone of positivist research, accepted as being the guarantor of reliable knowledge production. Of course, objectivity has been subject to rigorous feminist critique and shown to be a concept which purportedly signifies neutrality, when in fact it masks questions of unequal gender and class power relations and works to reinforce male dominance (Harding, 1991; Olesen, 2000; Schwandt, 2000). In this sense, we can see the trend towards researcher reflexivity as a politically motivated response to this. Social construction theories have argued that knowledge of the subject existing outside of its constitution in language was impossible. Certainly, a good deal of television research has, in recent years, focussed on "talk", and a number of studies have looked at the way classed and gendered identities are constructed in talk about TV (see Wood, 2009; Wood et al., 2008; Tolson, 2011, 2006). Similarly, emotions are often viewed as "performed" or discursive constructions, shaped by and expressions of socio-cultural and historical forces (Rustin, 2009). Interview data is also said to be a linguistic construction shaped by institutional discourses and confirmed by social consensus, suggesting that analysis be restricted to the social and discursive significance of what is said. However, the implication of feelings and ideas moving between people (sometimes without the conscious awareness of the individual), suggests that the study of communication in interviews, and the analysis of the data produced, cannot just be limited to what is said, as found in traditional qualitative interviewing reasoning.

Positivist research assumes a cognitive, rational individual, conscious of their actions and whose responses are reliable and under the control of the subject. Whilst critical researchers are alert to the constraints placed by discourse and social power on the individual, in most qualitative interview-based research, it is presumed that what the subject tells the researcher is largely truthful, transparent and coherent (Walkerdine et al., 2001; Hollway and Jefferson, 2010). The basis of Hollway and Jefferson's critique of standard interviewing is based on what they call the "transparent account problem" proposing that it is a common mistake of qualitative researchers to assume that their interviewees know what their experience means and can present a logical, unproblematic and transparent account to a researcher, who is then able to make sense of it (Hollway and Jefferson, 2000, pp. 56–59, see also Wengraf, 2001). There can be difficulties in comprehension for both parties in an interview (see Chapter 6). Usually the researcher is positioned as one who is able to maintain distance and is considered able to interpret the data accurately with their own subjectivity held in abeyance. Generally, there is no consideration of unconscious processes. If we acknowledge, however, that both researcher and researched have a unique biography and a set of internal object relations, which impact on the way reality is experienced, and if defences such as splitting, denial and projection are mobilised, then this makes the idea of a calm, controlled, objective research situation seem much less stable and open to control (Walkerdine, 1986).

Understanding the workings of the inner world, and the role played by phantasy, means psychoanalysis is able to offer reasons why interviewees might not tell factually accurate, coherent or realistic stories. Whilst the unconscious is not tangible, it can be observed through "symptoms" of the "psychopathology of everyday life": emotions, affective states, dreams, fantasies and even jokes (Freud, 1901). These quotidian presentations of the unconscious, such as understanding factual inaccuracies as parapraxes (an unintentional utterance thought to reveal an unconscious motive or wish, which if acknowledged might provoke unconscious anxiety in the subject), will form part of the discussion of the case studies in Chapters 3–6. Such inaccuracies might have been discounted in some traditional approaches to research. Rather than being discounted, psychoanalysis might understand such mistakes as part of the interviewees' unconscious concerns which drive the interview agenda, but in a far from coherent and logical way.

It is not just the interviewees' unconscious that is our concern. We must take account of the researcher's subjectivity and the reality of their phantasies, anxieties and desires. Feminism's consideration of the personal has forced the self and experience onto the agenda. In media audience research, as we will see, this largely focussed on the social characteristics of the researcher. More recently, psychosocial researchers utilising psychoanalysis have encouraged serious engagement with researcher subjectivity as a resource to facilitate knowledge (Hunt, 1989; Walkerdine, 1986, 1997; Walkerdine et al., 2001; Hollway and Jefferson, 2010; Edwards et al., 2006). The researcher's subjective responses, their thoughts and feelings, are not slippery variables that must be discounted, but are an example of the way in which the researcher's self is a "primary instrument of inquiry" (Hunt, 1989). The idea that research subjects might not understand their own responses applies equally to researchers, who are not necessarily in a privileged position to understand their own reactions to what is being thought and said in interviews. Psychoanalysis does, at least, provide concepts and theories that researchers can use to help recognise these limitations in a way that the naivety of the concept of objectivity does not. This will be explored in more detail in the final section of this chapter and also in Chapter 6.

As well as the idea of researcher reflexivity, it is necessary to consider unconscious communication. In answer to those who are sceptical about psychoanalysis, empirical evidence has emerged from research in neuroscience that degrees of unconscious communication and projective processes are present in all our interactions with other subjects (Schore, 1994, 2010).[3] As noted in the introductory chapter, transference is not a phenomenon of the consulting room but something "non-pathological" and present in "everyday social relations" (Andersen and Berk, 1998, p. 81). Transference, counter-transference and projective identification are examples of mechanisms, often used defensively, involving the transfer of feelings between people. For instance, difficult feelings can be projected outwards towards a specific object in response to anxiety (Klein, 1946). Significantly, this then induces feeling states in the recipient "that are congruent with the ejected feelings" (Ogden, 1990, p. 25) and the "receiver" can come to know how the unconscious phantasies of the "transmitter" *feel* (Schore, 2003, p. 59). Bion's (1962) "container-contained" model is an account of the way in which projective processes can be used in a more benign form in processes of "attunement". The infant "learns from experience" as it projects difficult feelings and sensations into the mother who is able to contain and mollify these feelings and return them to the child in a form that can be tolerated (Segal, 1992, p. 122). The mother initially attunes herself to the child, and eventually they both learn to adapt and respond to the other's emotional and affective state. From these early emotional experiences, the child deduces "the models and concepts he will use as hypotheses in his contact with internal and external reality" (Grinberg et al., 1993, p. 100), thereby having a personalised emotional template towards the perception of reality. Whilst this is an important way of thinking about communication, it is also linked to the way in which emotional experience represents a form of knowledge acquisition. "Knowing" is the activity through which the subject becomes "aware of the emotional experience and can abstract from it a formulation that represents this experience in a relatively adequate way" (Grinberg et al., 1993, p. 100). These abstracted elements are used in processes of understanding the other subject or part of the self and, it seems clear that this is something crucial for research/researchers.

The researcher's mental experience, thought of as dynamic and involving intra-psychic processes and responses to discourse and demands of the social, must be considered in research as it is said to "mediate" their "understanding of the cultural and psychological world of subjects" (Hunt, 1989, p. 13). In practical terms this involves keeping a research diary that records thoughts, feelings, emotional responses and fantasies over time, as well as not rushing to interpret the material but "living" with it and being open to how the material might speak in multifarious affective ways to the researcher. For instance, whist working on Sophie's data (who is someone who adores and collects soft toys, see Chapter 4) I found myself at a car boot sale and realised that I was holding and stroking a *Beanie Bear*™ and feeling how *lovely* it was. I realised that this was a response to Sophie's data as I was able to emotionally identify with her. This also suggested she had been able to communicate her experience to me and that I was able to respond to it and, in part, live it, whilst recognising that it was not my own experience. I was able to enter into an understanding of how it feels to connect with a consumer object of the kind I would not have ever considered purchasing or becoming attached to. This openness to feelings and responses associated with the research experience can be used in forms of triangulation with other forms of evidence, such as biographical details or parapraxes present in the research data.

Recognition and acknowledgement of difficult thoughts and feelings, such as uncertainty or aggression, can help researchers in their analysis of data as it is possible that the presence of unconscious anxiety might precipitate the emergence of defences, such as denial of something in the data that challenges the researcher's assumptions. Researchers can be alert to their responses, and it is within human capacity for subjects to differentiate between what is *our* feeling and experience and what *belongs* to the other. It is possible to learn from Sophie's love of soft toys by temporarily identifying and engaging in the experience, recognising her for who she is. Equally, as the case study in the final sections shows, researchers must also be aware that often they have desires for how they want their research subjects to be, which is in conflict with the research subjects' views of themselves. It is only through an acknowledgement of the researchers' responses that it is possible to work out how the subjectivities of both parties are impacting on each other (see also Chapter 6). Significantly, the idea that communication and knowledge should be understood as something affective, and that relationships with objects are both cognitive and emotional, is something new in audience studies. In fact, it would be rare to find a communication or media studies textbook that offers a theory of unconscious communication and media messages placed alongside traditional sender-receiver models. This would also extend to research methods guides in these fields.

I will now outline the method and use of case studies, describe some of the theoretical underpinnings of the approach and give examples of how the ideas described above translate practically.

Free association narrative and biographical interviewing

The opening chapter introduced the audience study and sample rationale on which later chapters are based. The study was based on twelve in-depth interviews and, with one interviewee, a series of letters. The respondents' names have been changed and personal details removed to ensure anonymity. I had not planned to use letters specifically, but Bill (Chapter 6) had asked if we could correspond by letter. Each interview lasted for approximately two and a half hours. Long interviews were preferred as a means of building a relationship with the research participant and to allow for reflection on intersubjective processes which were explored through research diary entries after the interviews and as the data was coded. As the research was interested in biography, allowing the participants the time they needed to discuss favourites and anything else they wanted to talk about was an important aspect of the interviews (see also endnote 5 Introduction).

To recap, the study is a qualitative, in-depth, biographically influenced, interview-based study, using a psychoanalytically informed method: Free Association Narrative Interviewing (FANI) method (Hollway and Jefferson, 2000/2010). In-depth has two meanings with regard to this piece of audience research. The first is the wish to gain a greater understanding of the viewing experience with particular attention being paid to the significance of the viewer's biography. The second is the desire to move beyond the surface or manifest content of discourse, (whilst recognising its role as a mediating force), to explore latent processes with a view to developing the use of psychoanalytic concepts in research. For this reason, a small sample, based on a case-study design, was chosen with the aim of achieving depth rather than the breadth associated with large scale, generalisable research. Long interviews were also preferred as a means of building a relationship with the research participants and enabling reflection on researcher subjectivity.

Qualitative research interviewing tends to work the surface level, in that it assumes that an interview is a transparent window on psychological or social realities (Wengraf, 2001) and that the meaning of what is told to an interviewer is transparent to both parties, even when, as I have explained, subjects are sometimes puzzled by their own actions and statements. FANI is based on the idea that research subjects unconsciously defend against difficult and conflictual thoughts, feelings and ideas which, in conventional interviewing, produces responses which miss the emotional and experiential significance for the interviewee. The method draws on techniques from biographical and narrative research where the interviewee is asked open-ended questions designed to elicit biographically significant narratives to gain access to the way in which this particular individual experiences the research topic in question (see Hollway and Jefferson, 2000/2010; Wengraf, 2001).

The role of theory, in the design of the research project, draws pragmatically on two traditions: inductivist and deductivist models. My inductivist strand draws pragmatically on grounded theory. The data is examined for relevant facts, and the most appropriate theoretical concepts and interpretive frameworks emerge out of the data (Glaser and Strauss, 1967; Corbin and Strauss, 2008; Pace, 2012). The research was also designed with psychoanalytic theories in mind in a flexible deductivist vein to explore the utility of psychoanalytic concepts. The aim was to let appropriate concepts emerge from the data, rather than starting with a strict a set of concepts that I wanted to examine.

The FANI method can be seen as a realistic, pragmatic and democratic method. The aim of the method, to generate narratives, is based on the insight that narratives are "meaning frames" that help make experience meaningful, even if this is unconscious meaning (Nelson-Jones, 2001; McLeod, 1997; Schafer, 1981; Spence, 1982). Gestalt psychology/psychotherapy proposes that perception is influenced by the character of the inner world and that humans do not perceive things in isolation but that they organise and configure experience into significant "wholes" (Nelson-Jones, 2001; O'Shaughnessy, 1994). Narratives are important expressions of experience. They are not analysed as isolated stories but as part of the whole data entity, which also includes the researcher's feelings and responses as previously explained.

In conventional qualitative data interview analysis, only data that appears directly relevant to the topic is considered. Information that doesn't fit is often discarded, as we will see in the final section of this chapter. In this case, only material that directly relates to favourites or films/programmes would be considered relevant. Nevertheless, this does not tell us where the concept of favourites fits with the interviewees' wider experiences and current psychosocial concerns, which shape the way in which the concept of favourites is meaningful for the interviewee. For instance, Chapter 5 demonstrates the significance of key moments in Mary and Sue's childhoods concerned with the birth of siblings. This theme did not emerge from logical and coherent sets of statements. In fact, many statements in the interviews (and letters) seemed on the surface to be off topic, irrelevant and illogical. It is only in viewing the interview as a whole, that the material could be understood as having any sort of patterning or theme (see Chapters 3–6).

This is also in the spirit of Freud's free association, which is another feature of the method. This is a clinical technique where the patient starts with a dream fragment, a memory or whatever thought comes to mind. The patient is then encouraged to give voice to any subsequent thoughts which enter the mind. These can be words, numbers, images or any kind of idea. This translates in FANI into an approach that does not discount the illogical and considers the narratives interviewees tell as meaningful even if they do not directly relate to the topic. The whole data is taken into account rather than just those items which follow a logical order. Free Association was designed to both accommodate the illogical and to overcome problems associated with accuracy and the recall of events in analysis (Freud, 1900; Freeman, 1989). Freud used free association to counteract the claim that the unconscious could not be observed. The epistemological claim is that free association provides a link between the conscious and unconscious worlds. It prevents the process of interpretation being arbitrary because free associations are never completely "free" (Freeman, 1989). Freud thought that the original idea that a clinical session started with an 'idea' (translated in this research to "favourite texts") and this starting point idea was important as it prevented "irrelevant" material emerging, maintaining the link with the "significant" subject matter (Freeman, 1989).[4] Free association also points towards the nature of the unconscious, in that seemingly opposing ideas and images easily co-exist. The method helps the interviewer make sense of material that is "riotously anti-contextual" (Bollas, 1995, p. 12). The unconscious ignores time, for instance, as moments from early childhood are represented alongside contemporary events. Hermeneutic and life history approaches to narrative analysis often take a developmental form, organising experience around past, present and future. However, as the case studies show, sense-making is never purely rational, as the work of narratives has unconscious features. Thus, the presentation, form and structure of narratives can be viewed as vehicles through which the unconscious and emotional concerns can unknowingly and unwittingly emerge.

Free association informs the research design practically, using a particular form of open questioning, to facilitate the production of emotionally meaningful narratives and other trains of thought that can elucidate thematic wholes. In practical terms, this makes the request "Can you tell me about anything that comes to mind about your favourite films or programmes...?" productive, as it allows the interviewee to discuss particular incidents that have emotional significance, without attempting to pin the interviewee down to talking exclusively about a film or programme. This was based on a critique of the standard question and answer format associated with interviewing, which relies on "Why?" questions (Mishler, 1986; Wengraf, 2001) which are said to close down potentially fruitful avenues, as they force interviewees to *intellectualise*, which can be used defensively to allay unconscious anxiety. At the same time, they force information into the straitjacket of the research topic by making information fit. The open question and narrative generating method clearly facilitates forms of free association as interviewees often said, "I don't know why I thought of this, but I must tell you..." or "This is nothing to do with this film but this just came to mind...". The information formed a link between the text and something that was significant for the viewer. Thus open-ended questions which allowed interviewees to set the agenda by asking them to talk about any favourite film or programme, scene or character that came to mind were used, along with other methods such as watching significant clips with interviewees and asking the interviewee to talk about the clip as they watched[5] (see Chapter 4). Letting the interviewee talk was paramount. Responses were then followed up according to the interviewees' logic and ordering of events, rather than a research agenda.[6]

What kind of evidence do biographical narratives provide then? In the humanities and social sciences, narratives are said to point to the construction of identity, through talk and storytelling, which is a form of social and interpersonal action (Harre and Gillett 1994; Polkinghorne, 1995). In keeping with a psychosocial perspective on identity, a discursive analysis can be coupled with the idea that narratives are specific instances of events that have emotional significance for the speaker. They are creative expressions of inner processes as well as social constructions. For example, Chapter 3 studies Daniel and Chris who are both aware that watching television has negative cultural meanings, which makes them *feel* uncomfortable and anxious. This is double-edged, however, as the method inheres on its reliance on language to elucidate the intersubjective, embodied, communicative dynamic. Acknowledging this limitation, I propose that free association narratives can be triangulated with a number of forms of evidence such as textual analysis, history and researcher subjectivity.

If the triangulation of narratives with the text is to be used to meet psychosocial aims, then a different view of "reading" texts is needed. O'Shaughnessy (1994, p. 44), in a valuable, but not widely acknowledged piece, argues that where there is strong biographical identification with a favourite film, the viewer's readings of the text reflect emotional concerns. Often, the viewer's readings do not make sense with regard to anything that a formal evaluation of the encoded or constructed meanings within the text might suggest. O'Shaughnessy calls these "textual mis-readings" (ibid., p. 45), which is a rather negative way of pointing to something that is highly relevant; that emotional concerns can sometimes overdetermine the interpretation of a text. The idea of a "dominant" or "preferred" reading (Hall, 1980) of a text is useful, in that it provides a point from which interviewees' narratives about the text can be compared. For instance, Sophie (see Chapter 4) finds a scene in *Castaway* very distressing. Whilst the scene was designed to be emotive, my reading of this scene suggested that her highly distressed response was in excess of anything that would realistically be generated by the events on screen. O'Shaughnessy (op. cit.) argues that this is symptomatic of readings where biographical identification, shaped by emotional difficulties, is involved. To triangulate my reading, I used the *Rate It* website, popular with people of Sophie's age, and confirmed that her response was not typical. In fact, the majority of the responses found it rather corny. When Sophie's response was linked to her biography, however, and set against the back-drop of historical and social change, then the distressing reading is understandable (see Chapter 4).

Case studies, theory and troublesome audience research

Individual case studies or vignettes have traditionally been the preferred method for the presentation of clinical psychoanalytic data. It was felt that case studies would be appropriate for this study, which was not looking for patterning across cases which would require a larger sample. As a social science research method, the case study has "long been stereotyped as a weak sibling" (Yin, 1994, p. xiii, see also Yin, 2012). They are often accused of being purely descriptive, the detail is often too great, and theory and explanation become lost in the detail. But the most salient criticism applied to case study and small sample research concerns their inability to be replicated and generalised. The latter ambitions have dominated much positivist and qualitative research. Case studies and small sample research cannot be generalised *statistically* to "populations or universes" (Yin, 1994), but can be used, in keeping with my aims, to generalise theoretically and conceptually, as long as the researcher is aware of constraints (Dreher, 2000). Moreover, Dreher argues for "systematic" study of psychoanalytic concepts and promotes setting up empirical work outside of the clinical situation to explore the appropriateness of concepts for different situations. So, rather than being based on statistics and probability, it is possible to make inferences about the applicability of theoretical concepts to other situations, both similar and different, whilst noting any particular constraints that might be in operation. As later chapters will show, important cross-case findings and conceptual developments were made, whilst remaining alert to the constraints placed upon research design and findings associated with differing contexts, social and cultural locations.

It will become apparent that each case study contains diverse detail which is unique to the individual. However, I will demonstrate in the following chapters that this diverse detail (context, biographical experiences, position in family, occupation, to name but a few), whilst relevant and specific to each case, also allows for conceptual and theoretical generalisation beyond this study. The in-depth case study approach adopted demonstrates the rich potential of psychoanalytic concepts and theory for audience research and also the role of the researcher as an instrument of knowledge. Even in a small sample, there were unexpected findings that point to the need to re-think important ideas, particularly around the concept of identification and media texts, for instance: intergenerational identifications, siblings and identification, and the role of the internal objects in family viewing.

Audience research

I want to bring together, in this section, a number of the themes I have discussed above, in particular researcher reflexivity and the use of psychoanalysis in interpretation in audience research, which will then be followed by a discussion of ethical issues. I will offer a critical reappraisal of a much cited piece of audience research which reflects precisely on the issues and tensions I have raised in this chapter.

Within audience research, psychology has been frowned upon and has been criticised for pathologising audiences and promoting the idea of passivity and "media effects" (Walkerdine and Blackman, 2001). Thus, in the case of interpretation in audience research, an important idea has been that the only really "safe" interpretations are those which reject psychological paradigms and focus purely on understanding the social implications. In keeping with this and in response to the demands for greater reflexivity, researchers have been encouraged to explore and challenge the power relations of the research dyad which have become dominant issues in terms of ethics as well as reflexive practice. Seiter and Borchers' (1990) *Making Distinctions: Case Study of a Troubling Interview* is a piece that explores how social and power differences can hinder communication. This is because such differences may inhibit researchers from asking appropriate questions or listening to answers leading to mis-readings of data or even making interpretation impossible (Johnson et al., 2004). Whilst this is, now, quite a dated study, the issues it raises for sociological audience research remain unresolved. Another reason for using this chapter is that it *deals explicitly with difficult feelings* for the researchers, especially around the *failure* to produce data in a study on soap opera audiences. Significantly, it also provides a transcript to allow the reader to interpret the data for themselves, which is rare. Researchers are fond of stating that data can always be re-interpreted by others, but in actuality this is impossible without a full transcript. Thus the case provides a very useful opportunity to do just this, which will also show the relevance of a psychoanalytic perspective.

The chapter discusses the researchers' responses in an interview situation towards two male, retired housemates who responded to a newspaper advert recruiting interviewees who watch soaps. On arriving for the interview the researchers found that they became angry when the men began by claiming they neither viewed nor liked soaps! The researchers were forced to acknowledge the presence of strong emotional reactions of "annoyance", "antagonism" and "irritation" towards the interviewees. They recognised that these feelings were associated with the failure to generate usable data. In fact, they struggled to make sense of data that "just did not fit". Whilst the men would not stay on the topic of soaps, the researchers were able to identify that, in the range of topics they discussed, a theme of "self-acquired" knowledge emerged. My reading noted also that the interview also contained numerous, seemingly irrelevant and unconnected memories and biographical stories. In the end, the researchers concluded that the interview had been a "waste of time" and that they had learnt nothing about soap audiences. They also concluded that their interviewees were "crackpots". The researchers then wrote this up as a discussion of the differences in social power between the working-class men and the academics (one who was female) and argued, in a limited way, that these differences worked against knowledge production. In effect, the tensions between them, because of their power differences, stopped them from communicating.

When social and cultural research is presented, it is commonplace to find some form of a statement about the social categories of the researcher, for instance, white, female and middle class. The reader is then supposed to be able to make a reasonably accurate appraisal about the way in which this impinges on the research design and knowledge produced. In this honest and frank piece, Seiter (1990) acknowledged that awareness of social differences meant that she was *often* troubled by difficult feelings generated for her in this and other interviews. She also acknowledged that attachments to particular political perspectives on gender and class led her to focus on *social* factors, at the expense of exploring what her strong feelings might mean. To counter this, the researchers, at this point really thinking ahead of their time, designed an unstructured interview which would take cues from "digressions" and be receptive to "unanticipated areas of discussion, with the psychoanalytic scenario in the back of our minds" (Seiter, 1990, p. 60). However, it is clear from the piece that the process of knowledge acquisition was generating anxiety.

Psychoanalysis has a fundamental aim to explain "Why we feel deeply about certain things, certain experiences and people" (Chodorow,

1999, p. 3). For Hunt (1989), the presence of strong negative feelings in a research situation indicates unconscious defences against anxiety. Seiter realised that the strong feelings that she had experienced could not be explained fully by the social-cultural model she uses. Despite this, she then omits discussion of psychoanalysis in favour of the assertion "cultural studies must focus on the differences in class and cultural capital" as it is the *social processes* that "characterise the relationship between the academic and the subject of audience studies" (Seiter, 1990, p. 61). Seiter concludes that strong feelings can only be understood as symptomatic of issues of power resulting from hierarchies of class and gender. It is these hierarchies that work against the production of knowledge. Seiter and Borchers's self-critique is limited to an assertion that the class and educational differences between the researchers and interviewees produced the antagonism. This is a convincing argument, but I take issue with the idea that it is *only* social processes that can be analysed and to do so is to limit the possibilities for knowledge acquisition.

One of the ways I think we can understand the interview is through an idea introduced previously; "omnipotence". Winnicott (1971) and Bion (1962) have suggested that "omnipotent thought" must be relinquished if reality is to be objectively perceived. For Bion, all experience falls into three groups of emotion: Love (L), Hate (H) and Knowledge (K) known as "links". Link describes the emotional experience present "when two people or two parts of a person are related to each other". K is particularly relevant to the research situation as it dominates when there is a "Subject that tries to know an object and an object that can be known" (Grinberg et al., 1993, p. 99). The K link has a particular "shade" of its own and is characterised by painful feelings of frustration inherent in the subject's ability to "know anything". For knowledge acquisition the subject must learn to tolerate doubt.

In the container-contained model, knowledge acquisition results from the modification of pain and frustration in the K link (Bion, 1962). In infancy the knowledge, which the child attains learning through emotional experience, is built upon and used for further discoveries. However, knowledge *possession* can also be used defensively to *avoid* pain and frustration. This is associated with the omnipotent parts of the personality. There is a negative instance of the K link, minus K (–K) where the container-contained model is reversed and "Where meaning and emotion are actively denuded of vitality and sense so that discovery and development become impossible" (Grinberg et al., 1993, p. 100). In essence Seiter and Borchers tried to possess and control knowledge, rather than acquire it, by being open to the emotional and subjective aspects of the experience, and this helps explain Seiter's frustration and anger toward her interviewees.

I want to draw attention to the presence of unconscious anxiety and its effects on the interview. Seiter identified the themes of self-taught learning and knowledge that were present in the data, and I think that that the researchers accurately interpreted this in relation to questions of power and social hierarchies. Clearly there are significant differences that cannot be ignored between academic researchers socialised in legitimate culture and the working-class men whose knowledge was self-acquired and unconventional. This would certainly generate anxiety for the interviewees. For the researchers, however, there is equal cause for anxiety, given the reality of economic and time constraints placed on academics, and the pressure to produce publishable research, it is likely this made the interview situation equally anxiety-provoking for the researchers. The nature of the market forces that apply to academic research production, I would argue, creates the grounds for omnipotent thought and predisposes researchers to approach knowledge production through –K, negating the emotional components of research.

Also, there are questions around the researchers' investments in established theoretical and political positions which they brought to the interview. The researchers were irritated because, unlike other interviewees who were soap "literate", these men "couldn't even reproduce critical categories common to the *TV Guide*" (Seiter and Borchers, 1990, p. 62). Within Cultural Studies there had long been a celebration of inverse forms of cultural capital based on oppositional, bottom-up knowledge generated by fan cultures. Walkerdine and Blackman (2001) have critiqued this trend, arguing some academics' approach to research is often structured by fantasies about the working class, particularly concerning "resistance" and oppositional forms of interpretation. Had the interviewees been "soap literate", then the researchers' responses towards the men may have been more positive as this would demonstrate forms of "oppositional knowledge". This raises some thorny questions about the investment of some academics, in particular discourses and disciplinary influenced forms of knowledge (McGuigan, 1992, p. 4), and the concomitant defensive responses when these positions are challenged.

This adherence to particular social and cultural models, as Seiter acknowledges, found that the researchers were in a rather strange position in relation to their interviewees, which can be explored through the psychoanalytic model they reject. There is a degree of splitting present in the chapter in relation to "knowledge", (which also corresponds to social power, as Seiter notes). The "Self-taught" knowledge, which the interviewees wanted to display to demonstrate their intellect, was the source of much "irritation" for the researchers: "Mr H was trying to display his knowledge about a topic he (erroneously) assumed we would be interested in .... We were more receptive to Mr D when he returned to the topic of soap operas" (1990, p. 63). They were only interested in non-legitimate forms of knowledge when they met with the researcher's interests of soap opera and popular culture. The researchers were faced with feelings of pain, frustration and threats to forms of omnipotence associated with the need to produce knowledge predisposing the situation to one of knowledge possession rather than knowledge acquisition as described by Bion.

The interviewees were actually very open and presented the researchers with personal experiences and thoughts, but for the researchers these anecdotes "got in the way" of the knowledge that they were interested in. Much to their annoyance, the men spoke about their interests and tales from their past. This was the information that "didn't fit". Here, the experiences the men projected to communicate with the researchers were rejected. There were only certain types of knowledge the researchers wanted to know about. This indicated the presence of –K link, where the interviewees' experiences were internalised in a way that "strips and denudes of meaning" (Hinshelwood, 1989, p. 299). For Hinshelwood, this instance of –K is "one version of a harsh super ego" seen in the "denigrating" of the object (1991, p. 299) as the interviewees are labelled by the researchers as "crackpots" (Seiter, 1990, p. 65), which might also be seen as an act of aggression on the part of the researchers. Clearly, this interview was overflowing with seemingly irrational, powerful and conflictual feelings for both parties, accompanied by sustained efforts by the researchers to bracket and control such feelings. I will now describe an approach that would enable researchers to see these difficult feelings as a resource from which knowledge can emerge.

The tensions around "knowledge", which I would view as a series of projections shaped by fears and phantasies about the other, are also symptomatic of the men's internal conflicts associated with *ageing*. Confronted with the negative feelings of the researchers, Mr D and Mr H refuse a dependent stance and in part the idea of self-acquired knowledge; pointing to agency and activity represents this.

Under capitalism, social hierarchies can be structured through binary oppositions which involve evaluative discourses, with one side of the pairing being valued and the other often denigrated. Layton (2002a, b) associates this with "Normative unconscious processes" that promote forms of splitting. For instance, autonomy is valued whereas dependency is denigrated because it is associated with groups with less value. These groups can also be objects of hatred, such as women, non-white people or immigrants, for example. However, dependency and lack of autonomy often become reality for older people. Thus, in this study, it can be argued that defensively the men promote their autonomy through discourses of self-learning. In this process, however, they produce a sexist discourse, which adds fire to Seiter's feminist antagonism. It would seem that Seiter's power and her argument is confirmed, showing how language plays an important role in this; the men adopt processes that seek to undermine researcher authority, but at the same time the researchers "confirm" their intellectual superiority and power, by labelling the men as "crackpots". It is clear that knowledge acquisition, and its value and meaning in relation to other hierarchical discourses, emerged from the anxieties and fears of both parties which circulated consciously and unconsciously between both parties during the interview. It is worth investigating the intersubjective processes further to see what can be learnt about research but also whether anything was communicated that might be used towards the researcher's aim of producing knowledge about audiences and the soap opera genre.

Mr H told a story about "anger" towards the television "box". The story concerned Elvis Presley who famously fired a shotgun at a television because he was so angry at the content. It was, in fact, this story that prompted the researchers to conclude the men were "crackpots". This illustrates on one level an extreme case of the transitional functioning of television and its ability to withstand abuse (Silverstone, 1994). However, in the light of the difficult unconscious dynamics of the interview, it is likely that that the frustration and anxiety Mr H was feeling towards the researchers, which in turn was a response to *their* feelings of frustration and anger, meant that the television in this story could also symbolically represent the researchers. Thus, indicating his unconscious phantasy of what he would like to do to the researchers and others like them. Walkerdine's (1993, 1997) work on television has pointed to the way in which the television and its content can be used as part of a process of relay through which difficult feelings between interviewees and researchers can circulate. This kind of exchange, structured as it was by the category of "knowledge", illustrates the idea that cultural capital can generate unconscious anxiety and phantasy responses (see also Chapter 3).

Standard research practice in the social sciences promotes *empathy* as essential for developing effective research relationships. It is often unclear what empathy means. Bion (1962) understands empathy as a product of communication that is based on projective processes between people. Omnipotence, as a response to anxiety, and uncomfortable affective dynamics, such as those illustrated above, can prevent one person hearing or understanding what the other has to say. It is only when omnipotence is relinquished that the self can be open to receiving projections from the other and vice versa. The researchers were angry with the men for sharing memories and life stories. Researcher omnipotence and authority was threatened by the presumption, on the part of the interviewees, that the researchers and the men were interested in the same things. It is often assumed interviewees "give" and researchers "take". These men wanted something from the interviews that was different to the aims of the interviewers (see also Chapter 6). The attempt to deny interviewees' attachments worked against empathy and also hampered the collection of biographical information that they wanted to share which could have been used to further inform their analysis.

Seiter does not propose any theories of viewing or genre based on the interview, but I would argue that, taking the interview as a whole, we can see a number of reasons why soap operas are significant for the men. There is evidence that the interviewees identify with characters and narratives in the soaps. Despite displaying a dislike and antagonism towards soaps, on reading the transcript, it is apparent that the interviewees did in fact watch numerous soaps. Far from being television illiterate, they were able to comment on narrative, plot lines, characterisation and the difference between daytime and evening soaps. It is true to say their opinion sometimes appeared "sexist" and did not conform to the "pleasures of soaps" models found in some fan literature. Occasionally, the men indicated an ambivalent approach to soap viewing. However, it is still useful to see the interview as an example of the way objects, or in this case a genre, can be used as containers for self-states and projections (Bollas, 1987, 1992). Selves can be *dis-attached* to certain objects (Redman and Whitehouse-Hart, 2008) and attachments are not always pleasurable. As Mr D says, with reference to soap operas and echoing a theme found in this book, "instead of trying to make you feel better they make you feel worse" (Seiter, 1990, p. 73). As later chapters show, the things viewers find difficult are often as revealing as those that are found pleasurable.

From the transcript, it was apparent that the settings, narratives and characters discussed in the interview triggered chains of associations that took the men back to painful and difficult experiences and relationships in their past. In turn, these highlighted present distressing realisations associated with ageing. It is impossible to understand this interview without recognising the significance of age, but age viewed psychosocially. As younger, employed men, they would not have been in the position to watch as much television. The very act of viewing daytime soaps makes reference to their increasing age and change of circumstances. Mr D tells a story about suffering a heart attack and having to go to hospital. He tells the researchers about the way he kept thinking about *General Hospital* and how it all became "real" as he was "playing the part of a very sick man".

For those wanting to find traditional psychoanalytic themes in the "text", Mr D and Mr H summarise soaps in a language that indicated that the unconscious is found in characters who are full of "hate, envy and jealousy" (Seiter, 1990, p. 72; see also Modleski, 1982), but this is too narrow. Soaps focus on a number of themes: relationships, families and intergenerational issues. For these men, whose age appears to have confronted them with the reality of failed and difficult relationships, soaps have been able to offer a space in which these issues can be addressed, in keeping with Layton's idea of repetition scenario, where conflicts are revisited with the aim of finding forms of resolution and change. The men have now, to some extent, been "feminised" by their changed circumstances. This compounds the anxieties generated around dependency which Layton associates with a projection of fear and hatred towards "others" such as women (Layton, 2002b), which they would not have been exposed to in earlier life as working males.

At the end of the interview we learn that "knowledge" and soap opera have age and familial connotations. Mr H says:

> I do resent the point that they seem to be so intent on. In most soap operas dealing with juveniles and juvenile problems [ ] and it is always, always, the parents who come out looking bad. It is always the [ ] father's fault. Father never knows anything. He's a boob – an idiot. But the kids always come out looking perfect. No fault can be found with them. Only the parents.

Mr D expresses similar feelings when he says: "That's my theme song. *I know what it's like to be young but you don't know what it's like to be old.*" They both agree and sum up these difficult feelings by placing them into an *object*: "Orson Wells made a very well-known song, a forty-five with that title", which they both want to "send to their kids".

The emotional "work" of mourning around ageing can also be seen in the to and fro relationship between the past in the present, and the present on the past, that caused such frustrating fluctuations for the interviewers. Soaps have a mood, affect and rhythm that relates to emotion, attachment, family, life and change (Lopate, 1977) which make them particularly useful for the work of mourning in the latter stages of life. At the same time, their association with the home and "women's work" makes them both appealing and difficult for the men. For these reasons psychoanalytic perspectives enhance media research on "feminised" viewing.

Some concluding thoughts

This chapter has offered some new perspectives on method and methodology in audience research which draw on psychosocial and psychoanalytical insights, in particular the idea that communication can be improved by the recognition of unconscious and psychic processes. I have explored the idea that reality is known emotionally and affectively and, following on from this, that researcher reflexivity and use of self plays an important role in the production and interpretation of data. I have shown through evaluation of an established approach to audience research that it is possible to produce informed knowledge from a single case or small sample when psychosocial perspectives are applied. Equally, as this case study shows, researchers must also be aware that often they have desires for how they want their research subjects to be, which is in conflict with the research subject's view of themselves. It is only through an acknowledgement of the researcher's responses that it is possible to work out how the subjectivities of both parties influence the data produced in interviews and other forms of communication. I want to note that the concepts, ideas and techniques that I have utilised from the clinical setting are not in conflict with ethical research and, in fact, they offer a critically realist perspective on the dynamics of the research situation and a nuanced and pragmatic view of the subject and its relationship to reality, which is missing from established approaches in positivist and qualitative research. The next four chapters will develop these ideas further as they present case studies of audiences and favourites informed by these psychosocial insights.

3
Spending Too Much Time Watching TV?

A key question for media research has been to understand the way in which audiences respond to and use television, as well as its role in daily life. Favourites are often linked to the specific personalised practices of viewers which are the result of changes to the viewing landscape, caused by the development of time-shift, record-replay and digital technology. In some cases, these special texts cannot be fully divorced from the everyday practices and settings in which they are viewed. These technological developments have occurred against the backdrop of a range of wider social developments, such as the replacement of the nuclear family by a range of reconfigured relational and domestic structures.

In this chapter, I concentrate on one neglected, but significant and increasing group: single men who live alone.[1] I present an account of 21st century television domestic viewing and leisure time in the lives of two single men: Chris (29) and Daniel (36). New patterns of work, involving the often pressing demands of relocating for employment purposes, mean that the idea of spending one's entire working life in the community in which one grew up and with family in close proximity is now anachronistic. Chris and Daniel both hail from working-class backgrounds and have experienced a degree of social mobility. They each live alone and are engaged in earning a living and building their careers, which inevitably points to their current situation as a time of identity change and transition. Through the study of the complex relationship with viewing in this chapter, we can explore some of the anxieties and lived experiences of this demographic group. Psychosocial approaches allow us to examine the negotiation between external and internal conflicts that individuals experience when identities and life styles change.

It has been written that the 21st century is characterised by the use of technologies which allow people to lead new forms of "connected lives" (Elliott and Urry, 2010). The *Facebook* and *I-phone* generation are thought to be able to conduct emotional and affective lives successfully through the use of digital technologies (ibid.). But what about those people who live alone and are *not* logged on to or plugged into virtual and networked worlds? In contrast to popular views of single people's communicative lives, Chris and Daniel do not spend all their time in a digitalised world of socialising and friendships. In fact, they both experience some degree of social isolation from their families and friendship groups as a result of changing employment and life patterns associated with late modernity. Both men like to watch television as a means of filling leisure time when they are not working. They would be considered by audience researchers as active audiences creating their own unique viewing practices using digital, time shift and record-replay technology as the reader will see. Politically, the imperative has been to look for the active, critical audience, but this has contributed to a reluctance to accept a view that television viewing could in any way be conflictual or be involved in emotional difficulties for viewers. Psychoanalysis would see this idea of conflict as an inevitable aspect of everyday life. The issue of time and how to spend it is complex and can, for some like Chris and Daniel, weigh heavily on the psyche, in particular provoking forms of anxiety. It is possible to understand why this might be the case psychosocially, by briefly reviewing the way in which television has featured in prevailing discourses and popular thought.

On one hand, television is thought of as ordinary and quotidian, a bit of harmless entertainment, but on the other it has been a source of social anxiety with fears expressed around the moral and intellectual welfare of audiences. The nature of this anxiety is two-fold, revolving around both the activity of watching television and the content. Prevailing discourses, supported by medical and effects research, see television viewing as low brow and banal, with concerns frequently expressed that spending time watching television is a "problem" and "bad for you".[2] The epithet of the "couch potato" is even found in modern dictionaries and describes the characteristics of: "A person who spends much time sitting or lying down, usually watching television"[3] or a "lazy person whose recreation consists chiefly of watching television and videos".[4] Whilst this is clearly a term of abuse, it also captures something of the viewing practices and activities of the men featured in this chapter, as the reader will discover. Television discursively is a contradictory entity; both a harmless object and an object of anxiety. To spend too much time watching television is considered a marker of a lack of culture and idleness. These conflicting ideas about television permeate, and become a feature of, the viewing experience for the men in this study.

Discourses featuring television which can be negative and evaluative do not exist in isolation, intersecting for instance with other discourses that structure taste, gender and class in a complex and often anxiety-provoking mix. Both Chris and Daniel express anxiety about the amount of time spent watching TV. They are concerned that some of their viewing practices, such as repetitive and long sessions of time-shift viewing, are both "bad" and "sad". Both men enjoy watching television and films at home, but they worry about "time-wasting". It is a fact that both men have leisure time to fill, yet why should television be both a compelling and yet problematic activity for them? What is it about television and time that might provoke anxiety, and what does it mean for these men?

This chapter also addresses a point made in the introductory chapter that it is possible to revisit the uses and gratifications paradigm psychosocially, to explore the suggestions that television can be used to problem solve, with specific reference to what this might mean in relation to emotional issues and identity conflicts. It also addresses the idea that television meets a "need" to "escape" and "pass the time" (McQuail, 1987). This chapter will explore, in more depth, what these needs and conflicts might be, something previously not explained fully. In this chapter, I want to show how an understanding of viewing as anxiety provoking can illuminate historically specific psychosocial conflicts about identity. These conflicts arise out of the particular neo-liberal social organisation of late capitalism in which audiences exist and watch TV. As the chapter will show, television plays a significant but contradictory role in Chris and Daniel's lives, where the idea of uses and needs can only be understood as part of a series of complex psychosocial processes that constitute viewing.

Filling time

Daniel (36) is single and lives alone; he has never had a live-in partner but he has had girlfriends. He has lived alone, reluctantly, since his late twenties when it was decided that it was probably not appropriate that he, as a young professional, live at his parents' home forever. He has a mixture of male and female friends, many from work. However, Daniel explained that he was "not sure why" but increasingly he sees less and less of his friends and family. He was also concerned that he does not have many hobbies apart from watching television and films. Although he will occasionally watch football and cricket matches, even this has declined in recent years, as he gave up his football season ticket and replaced "live" supporting with a subscription package to a sport channel. Daniel likes films but rarely goes to the cinema, preferring to watch at home, because cinema is more "demanding" on his time. At home it is possible to turn a film off, and at the cinema he is committed to observing spectatorship protocol and sitting through the film. He likes to be able to pause or turn films off.

Daniel has favourites he returns to numerous times and programmes which he likes that he has recorded and watched "ten times or more". He "gets absorbed" and "lost" in watching television. His favourite films and programmes can be summarised broadly as including texts which involve time-travelling; texts which feature "nice guys"; "detectives"; and what he calls "will they, won't they?" narratives. Daniel explained that he usually identifies with characters "that are sort of good, because it's the way I was brought up". He links this to a rather heroic self-sacrificing image of himself which also incorporates some aspects of his fantasy of being a detective:

> A girl at work said to me the other day. "All you seem to be put on this planet for is to make people happy." I said "Is that a bad thing?" and she said, "No it's a good thing, but you make other people happy before you make yourself happy." I've always fancied myself as a bit of a detective. Righting wrongs. I'd like to be helping people that sort of thing.

For Daniel, "there is always something good to watch" on TV. He is a quintessential modern active viewer, creating his own user-flow through his Sky Plus box, video and DVD player. Daniel has adopted a practice of delayed gratification, frequently saving up entire series to view in one go at weekends. ("I Sky Plussed every episode and watched it from beginning to end, just getting up to pause it." This prompted him to say: "I'm very sad aren't I?") Daniel refers frequently to viewing as a child, with his parents and siblings, as "happy times", and his memories stand in sharp contrast to his contemporary solitary viewing environment. Television was associated with his first taste of power. He recalls that, as the youngest child, he usually "got his own way" when the family debated what to watch. Television viewing also featured at a key moment in Daniel's life when, as a teenager, he was presented with a unique opportunity for independence when his parents went on holiday and left him alone at home for the first time. He spent the entire week, in a darkened room, watching the television series *Twin Peaks* (1990, Dir. David Lynch). He explains:

> I got a week off work and it was the time before I had a mobile phone and then the phone wasn't likely to ring at home and I wasn't going to be disturbed. So literally I would get up, sit and watch and watch and watch until bedtime. Get up the next day, watch and watch and watch and that's what I did. ... You lose yourself in the plot.

It is possible to see something of the dreamlike quality Williams (1974) ascribed to the continual "flow" of images and sounds that characterises television (*Twin Peaks* is recognised as having a bizarre and surreal quality associated with dreams). It is interesting that at this moment, when Daniel had his first taste of independence, he made a choice to stay at home and use viewing to structure his time. Perhaps his position as youngest child in the family and the only boy, who was allowed to control both his parents and siblings' viewing, contributed to turning television into an evocative object of unconscious investment. In the *Twin Peaks* example, clearly the mix of viewing, getting lost in the plot and his sense of autonomy provided Daniel with something pleasurable, but he is also aware that his choice was different from the activities of most of his peer group and perceived as "odd". He recalls that people at work found his story of a week of *Twin Peaks* as strange, and colleagues asked, "Is he weird or something?" If at 19, television viewing stood in opposition to being with his parents, now it stands in opposition to work. As the extract below shows, despite work being what he describes as "torture", Daniel has made himself indispensable. Time is strictly divided into "pleasurable" and "non-pleasurable" aspects, corresponding to the typical work/leisure divide of modernity which sees work as the space of alienation, and leisure as the realm of freedom and creativity.

> I haven't had more than a day's holiday off work in more than two years. I do overtime on Saturdays. I normally leave for work at 6.30 a.m. and get home at 6.30 p.m. ... A good film or television programme, whatever makes it worth going to work, it's what you've treated yourself to do. You know you've *tortured* yourself for that many hours, so you think, right I'm gonna make the most of my viewing time.

Within this, he also recognises that he procrastinates and puts off jobs that he doesn't want to do, so even within his leisure time he uses watching television to ease the tedium of chores and is careful to plan out his time:

> Last Sunday I watched *Scream II* and *Scream III* and I was doing my ironing and that was what I planned for my Sunday. If I'm *ever* going to do my ironing I've got to have something I can watch.

It was here that Daniel told me that he "doesn't know why", but he is no longer taking holidays with, or seeing, friends as often as he used to. In fact, he is reluctant to take extended periods of time off and even works overtime on some Saturdays. This decision to spend less time with friends seems odd as Daniel is a sociable, personable and intelligent man, who clearly gets on well with his work colleagues and is good company. It is possible to see how both work and television viewing take up time. Watching TV and even doing overtime fill up the large chunks of time which holidays and weekends generate, particularly in response to his self-enforced isolation.

In an attempt to understand the uses of television for Daniel, it is useful at this point to work towards a psychosocial view of time. Zygmunt Bauman (1999, p. 3) has argued that the separation of work and leisure associated with modernity creates particular "attitudes" to time, noting in particular the future as a "target" towards which the individual "travels" in the hope of fulfilment, "towards another present distinct from" and "more desirable" than the current present. A characteristic of this quest, Bauman suggests, is that individuals attempt to *manipulate* time, and time is constantly evaluated between modalities of the present of "differing quality and varying value". Technological advances, for example Daniel's much used *Sky Plus* facility, take viewing out of the hands of the scheduler and offer the viewer immense power to control viewing and, in so doing, also to control time and thus restructure various presents.

In this work-leisure separation, "work becomes an activity that derives its value from what it is not: it prepares the ground for non-work of leisure time" (Bauman, 1999, p. 4). On the surface, Daniel's expression of work as "torture" might be perceived as a contemporary expression of the Marxist concept of alienation. However, free time is not, in contrast, some utopian space for this single man living alone who, for whatever reason, is spending more and more time alone; it is a space that must be filled. Whilst work might be deemed an ordeal, Daniel has made himself indispensable at work, rarely taking time off, and this feeds what Bauman (ibid.) calls an "opposing tendency" of "work for work's sake". Work as an annoyance and torment, however, provides Daniel with a justification for filling his leisure time with viewing and offers an opportunity for the manipulation of time which viewing provides. This pattern of work and viewing has become the rhythm of Daniel's life. It is not clear if this is by chance or design. It is, however, something that causes anxiety, as he explains that he is frequently worrying that he is a "sad case", indicating a degree of interpellation by negative viewing discourses where watching TV is wasted time. I want to explore this further in relation to some of Daniel's favourite texts.

Time is clearly something evocative for Daniel, as both time and time-travelling feature in a number of his favourite texts, including television programmes such as *Dr Who* (BBC), *Back to the Future* (1985, Dir. Robert Zemeckis), *Forest Gump* (1994, Dir. Robert Zemeckis) and *The Time Traveller's Wife* (2009, Dir. Robert Schwentke) and a film he first discovered on TV and now owns on DVD, *Simon Birch* (1998, Dir. Mark Johnson). Both *Forest Gump* and *Simon Birch* make him cry, and he is aware they serve a cathartic function for him, and he views when "I just want to let some emotion out." Both films are united by narratives which have a sense of time passing and "not everybody lives happily ever after" endings. They also feature characters who are "nice guys", and ordinary men who do something heroic which are important features of Daniel's self-identity.

Forest Gump explores American history, inserting the leading character, a simple, nice guy like Daniel, into the momentous events of thirty years of US history and making him a hero. *Simon Birch* is a film that would not generally be considered a man's film. It tells the story of a life-limited child and his attempts to circumvent this restriction by achieving goals and attempting to do something heroic before he dies. The narrative explores time: the limits of time; the achievement of goals and dreams; the idea of time as something that is "under threat" and the finality of death. The main protagonist is aware he is dying and, if he is to be a hero, then he cannot afford to procrastinate and put things off, and putting things off and delaying gratification is something Daniel is well aware of. During the interview, Daniel produced a copy of the DVD, and I was struck by the remarkable physical resemblance between Daniel and the actor who plays Simon Birch's friend and father figure, another "nice guy". This suggests that there might be some biographical events or questions around parent-child identification, loss and desire that make the film so evocative and emotional. The identification is intensified by this physical resemblance, which Daniel must have registered even if unconsciously. Here, it seems, is a text that brings together time and its use in a text which provides strong possibilities for identification.

If being a heroic nice guy is important for Daniel it is interesting to explore what happens to nice guys in his favourite texts. Invariably things do not always go well for them, however heroic they may be: "Forest Gump is a nice guy and he's not got a bad bone in his body and you want him to be happy. You know it ended quite sadly." In the case of *Simon Birch*:

> The whole story is very, very sad. Fifteen years ago I could have watched that and not shed a tear. Now I don't know why, it's like the flood gates are open and you think, "Oh God, that's really sad." Whether or not that's because you get older.

His reason for selecting fifteen years ago is not apparent, although retrospectively he may locate this time as a turning point or a moment when he made a significant life choice as narrative and life history research has shown (McLeod, 1997; Roberts, 2002). Certainly, two years previous to this, he differentiated himself from his peer group by choosing to spend a week watching television at home. As we get older, past and present are evaluated and the future takes on new meanings that are not always easy for the self to contemplate (Freeman, 1993; Bauman, 1999). Youthful optimism is often replaced by the reality that "happy endings" are not always possible. Daniel explained that he likes the idea of time-travelling as it offers the possibility of moving back and forth through our own histories and in fantasy of making changes to the past. He is less certain about going forward and knowing what the future holds as this is "frightening".

Whilst these texts are suggestive of the fears associated with wasting time, Daniel is aware that he spends a lot of time lying on the sofa watching TV, so it is interesting to explore what purpose this might serve. It is useful at this point then, to continue to develop a psychosocial understanding of time and viewing, by introducing some psychoanalytic perspectives.

Firstly, it is not insignificant that Daniel talked a lot about family viewing, where television had been an important object used for entertainment, communication and intimacy in childhood. (Even now when he telephones his parents they usually talk about "what's been on TV".) This makes television an "evocative object" (Bollas, 1992, 1995) which links Daniel with family and childhood, even if this is idealised and coloured by phantasy, in response to the solitary life he currently leads. Further to this, I want to explore the idea that television functions as a transitional object (Silverstone, 1994) which, like other transitional phenomena, can be internalised into psychic structures that, in health, one can draw on for "comfort" in anxiety-provoking situations (Grolnick, 1990). Is it possible then, that Daniel's personalised user-flow is designed to help manage anxiety generated by "time"?

Winnicott's (1971) ontological concept of "holding" is useful as it can be used to explore the "experience of being alive", the nature of being and its relationship to time. "Holding" refers to the means by which a "sense of continuity of being is sustained over time" (Ogden, 2004, p. 1350), which Winnicott calls "going on being" (1965b, p. 44). In earliest infancy, prior to the child becoming aware of itself as a subject, time is "other" to the infant. The world of clocks, schedules, routine and calendars is profoundly threatening at this stage, where what is "not me" is extremely disrupting to the child's continuity of being. One important function of maternal preoccupation sees the mother forgoing her needs, for sleep for example, to absorb the impact of man-made time as she attunes herself to the infant's rhythms and needs. The mother provides a setting in which the child can gather together the various "bits", which eventually allow the child to feel integrated. Living alone, for Daniel, was something he reluctantly chose to do, and this would require some adjustment. For the infant learning to be alone, time becomes a symbol or mark of the external world beyond the child's control but, at the same time, it is connected to the child's physical and psychological rhythms (Ogden, 2004, p. 1353). Time is other to the infant, but it copes with this as it learns to experience a continuity of self over time. Transitional objects can help the infant manage anxious times whilst its mother is away.

Thomas Ogden (1989) also pointed to a primitive sensory mode of experience concerned with the infant's relationship to "surfaces", such as the mother 's skin, the rhythm of her breathing, and changes in temperature at the skin surface which sustain the child in a "non-reflexive state of sensory 'going on being'" (ibid., p. 32). He named this experience "Autistic-contiguous" and linked sounds and shapes and surfaces to the most primitive forms of holding. In later life anxiety can be allayed through a process of allowing oneself to become moulded to surface and sensory environment (ibid., p. 33). This offers an explanation for why lying on the sofa in the dark watching the flickering lights of the television forms a holding function for Daniel, and also other individuals, as a response to living alone or purely as an account of the soothing and comforting aspects of watching TV. Television viewing also becomes a setting, which Daniel can experience as a holding environment, because it allows him, through his manipulation of time, some sense of control around the demands placed upon him by time in the social world. This explains some of the complex gratifications the young Daniel obtained from watching a series in one go. It provided a form of absorption of time and provided a structure that was outside the conventional uses of time, day or night, which provided a template for his management of anxiety around living alone and managing time. The following section continues to explore the idea of time further through an examination of identification and Daniel's favourite texts, but first it addresses the methodological issue of feelings in interviews.

Unconscious communication: "detectives" and "will they won't they?" narratives

Psychoanalysis would recognise feelings in interviews as evidence (Spezzano, 1993) which can be utilised to understand the effect of the researcher on the data produced. My presence as a researcher, but also a woman of a similar age to Daniel, who was interested in his life and what he had to say, inevitably has an effect on the data produced (see Chapters 2 and 6). This provides another layer of evidence against which readings can be triangulated. I want to offer some thoughts on the affect that was present in the interview space and in particular a feeling of sadness that was present in our unspoken communication.

Daniel seemed to enjoy both interviews. I sensed sadness at the end of each interview, and it was difficult to bring the interviews to a close. He maintained a link with me by insisting I borrowed the DVD of *Simon Birch*. We spoke for many hours and talked about vast numbers of texts. I felt I understood how being "lost" in viewing felt, as I became lost in the flow of his discussion as he moved from one text to another and another. At interview one, Daniel offered to make me a drink. I asked if he had anything decaffeinated and he looked surprised and apologised as he "didn't buy it". Instead I had a cup of tea which he served in a cup decorated with a football logo. I noticed he looked slightly embarrassed by this, as it seemed to confirm the lack of a feminine touch in his home. On my second visit, he made a point of saying he had "found" me a "better" cup, a lovely lilac, floral, feminine mug, and I was treated to a special instant frothy decaffeinated cappuccino, which he had obviously purchased specially. I found this thoughtful, and I had the feeling he was giving me a gift and paying attention to my femininity. I could not ignore the feelings that I had, that Daniel was lonely and appreciated the company and that he regretted not being in a relationship. These current concerns seemed to be borne out in the interview as they appeared to force particular memories, texts and readings to the forefront. This will be examined below through a discussion of "will they, won't they?" narratives such as Twin Peaks and Daniel's practices of both absorption and delayed gratification in viewing.

Twin Peaks

It is possible to see this text as capturing something of the concerns of the 19 year old Daniel and the Daniel approaching middle age. Daniel explained that he "identified" with detective *Dale Cooper* and also that he was fascinated by the potential relationship between *Cooper* and *Audrey*.

> Well I think every time *Audrey* came on screen you were more interested as she was a good lookin' girl and she was flaunting herself around sort of. And you were constantly thinking you know, you sort of knew people who were a bit like that, sort of. You can sort of empathise with her. I'm a lad and she's a girl and you can sort of think, I know *lads* [my emphasis] who are a bit like her. And you wanted her to get off with *Dale Cooper* because he was such a nice guy, not a straight dull boring one sort of thing. You know she needs to be, you want her to be happy, she was a nice, likable character in a seedy situation sort of thing. I think sort of, every time she was on. It's probably one of those things, and I've got some *spare time* [my emphasis] I will probably sit and watch it again. I can still remember a lot of the episodes. I think they did this thing where she put a cherry in her mouth and tied the stalk of the cherry into a knot or whatever.[5] You know as a bloke you think "bloody hell," you know (laughs).

The scene captures a number of evocative figures for Daniel, detectives, heroes and nice guys, as well as providing a dreamlike fantasy aesthetic mimicking the flow of his intensive viewing experience. The sequence in question is noted as sitting formally and aesthetically on the cusp of fantasy and reality, possibly also suggesting a certain transitional quality.

The sexual and desirous nature of the extract requires some examination. The discussion of the scene also includes some intermingling of fantasy and reality in identity work. There are some interesting parapraxes and projections in the extract. Daniel says he knows "*lads*" who are like *Audrey*, the slip indicating cross-gender identification perhaps but, equally, it might say something about his desire of her attributes of good looks, sexual attractiveness, her flirtatiousness and control. At the same time, the fantasy figure of the detective is associated with a particular form of successful masculinity, which to some extent Daniel can associate with his life style choices: lonely, intellectual and in control. It is no surprise, therefore, that he is able to project himself into this scene in a number of ways, not unsurprisingly wanting *Audrey* to "get off" with the *Agent Cooper*, who just happens to also be a "nice guy". Daniel then explains that some nice guys are "straight" and "boring", but *Agent Cooper* wasn't boring and "nice guys" can also be "exciting". I sensed this qualification was included as Daniel realises his life may be perceived as boring, but he disavows this in favour of his vision of himself as a nice guy who is also masculine, cool and sexually attractive. This scene would have been viewed initially when Daniel was a teenager likely to be dealing with feelings around his sexuality. "Popping the cherry" is a colloquial term for loss of virginity so it is possible this was a concern at that time or currently signifies a longing for a sexual relationship as a form of afterwardsness, as past and present concerns merge.

One very obvious question arises that if Daniel wants and needs a relationship why does he structure his life around practices that preclude possibilities for meeting potential partners? If relationships are difficult for Daniel for whatever reason, viewing allows Daniel to displace and make safe his desire for a heterosexual relationship onto on-screen couples. He also has an ambivalent position in relation to romance in plots, preferring a form of delayed gratification to narrative closure in "will they, won't they?" narratives. He gives this example of another time-travelling favourite, *Dr Who*:

> To me that whole thing (new *Dr* Who) succeeded because it revolved round a male and female character will they/won't they? relationship. I watched *Torchwood* the *Dr Who* spin-off. But they ruined that for me because the key female character ended up copping off with one of the other characters and I found that totally unbelievable. That's sort of ruined it for me. Thinking the whole thing is built on two main characters will they/won't they? and now she's gone off with somebody else, you're thinking, well that's sort of *broken that barrier* that *you sort of wanted to carry on week in week out.* [my emphasis]

The narrative uncertainty is exciting as it acts as a "barrier", with narrative closure shutting down an important phantasy space and interrupting the pleasant on-going "flow" it creates for Daniel. It is also interesting that he viewed these texts rather like romances and not principally as sci-fi action-adventure. He is clear that it is the week on week delayed gratification, which has also been theorised as a form of procrastination (Bauman, 1999) that he finds pleasurable. The procrastinator has a goal but does not want to reach the target because there is finality in reaching the goal. The pleasure is in thinking about the goal and planning how to get there. This is similar to the staging of desire in fantasy (Cowie, 1997) and also explains multiple cross-gendered identifications within the *Twin Peaks* scene. Procrastination is a form of ambivalence which frustrates the goal by continually putting off reaching that which is desired. This is not just an approach to texts, however, as it is associated with Daniel's life style practices.

Daniel gives another example, of watching international football on TV, usually associated with group viewing:

> It's like if England are on I automatically think "well I won't watch it, but then it will probably be a good game, and everybody else will watch it and be able to enjoy it, but if I watch it, it will be absolutely awful."

Here is a man who likes football and subscribes to a sports package on television yet the thought of viewing, what for many would be an un-missable event, provokes anxiety. There is ambivalence about the match. If he watches it, it will be rubbish: if he doesn't, he will miss something special. This is another example of a will they/won't they? ambivalent position. It also suggests that communal viewing events, which suggest social interaction, communication and intimacy, are anxiety provoking perhaps because they act as a reminder that he is spending more time alone. Thus, there is a tension between delaying gratification, procrastination and the practice of intense viewing, all of which are linked to the question of time.

Absorption

Like many of the characters in his favourite films, Daniel experiences time as something under threat. It is limited and divided between the "torture" of work and his leisure time which looms before him and which must be filled. This is not an easy prospect for a single man living alone. Object relations theories emphasise the value of creative living in emotional well-being. This begins in infancy and involves being open to the unique qualities of the world of objects, responding to desires, allowing oneself to become *immersed* and *absorbed* in objects, which Winnicott (1965a, 1971) associated with the freedom he assigned to the "true", rather than an adaptive "false" self. However, material needs and social demands make work indispensable and, therefore, place limits on freedom, which Daniel recognises and, as an active viewer, subverts by using technology to manipulate time. Daniel opens himself up to and absorbs himself in the sensory dreamlike qualities of television viewing, which can be seen as a creative act that should enrich the unconscious. Bollas (1995, p. 92) proposed, however, that creative living is not possible without unconscious freedom and it is impossible to achieve this creativity in any meaningful way without "variety", "difference" and "unique- ness." Daniel, however, has chosen to engage with a narrow colony of objects consistent with what has been called "conservative object use" (Bollas, 1995; Mann, 2002). Such objects are symptomatic of the barriers that some individuals may unwittingly construct, which hinder the development of openness but, as I am suggesting here, they are also an unconscious response to social reality. Thus, while Daniel is open to the viewing "object", his viewing is contradictory, as he expresses concerns about "not seeing people" and says that he spends "too much time watching TV".

Is it possible that Daniel's absorption in television might be an example of what Bollas (1995) has called a "preoccupied" or "passionate" person? Where the individual "conjures up a mental space into which he can bring all of his interests to the exclusion of all else" (ibid., p. 79). In passionate object relating, the experience of getting "hooked" or "lost" (as Daniel did at a significant biographical moment with *Twin Peaks*) is compelling and the subject is changed by the experience, providing scope for creative thought and living. However, where there is anxiety, perhaps as a result of a social situation such as living alone, the individual "conjures up an object of such interest in order to pre-occupy the potential space of a more liberated object use" and this "betrays the anxieties that generate such a move on the part of the ego" (ibid., p. 88).

This absorption, combined with what Freud (1914, p. 244) called a "cessation of interest in the outside world", would also suggest melancholia and might also account for the combination of loss and preoccupation with viewing. Freud noted that melancholia can form when an individual reacts to a loss, which can be an abstraction, such as the image of family viewing in childhood. Freud (ibid.) noted that "devotion" to mourning what is lost can absorb the person so much that it "leaves nothing for other purposes or interests" and sometimes the melancholic person might not even be fully sure what it is that has been lost. Like Daniel, they are both puzzled by their behaviours and the melancholic subject often lacks insight into how habits and practices became established.

It is not possible to know fully what motivated the initial decision, almost 20 years previously, to spend his first week of independence watching television continuously, but this powerful sensory experience undoubtedly had an impact on Daniel. At that time, Daniel found that television viewing met a range of desires affectively and intellectually, while being "held" in time, and serving transitional functions. At the same time, this retreat set in motion a gradual, but steady, move over the next two decades that sees Daniel socially isolated and with his time divided in a stark dichotomy between work and home. The latter constructed around what Bollas (1995) might see as a "restricted" set of objects and activities. It is possible that his activity is a response to manage unrecognised feelings of loss around the march of time, from a happier past toward an uncertain, and what he acknowledges is, a "frightening" future.

By filling his time with something he calls "pleasurable", which it might be more effective to think of as absorption, it is possible that he does not have to confront the anxiety and difficulties associated with venturing into the world and trying to make relationships and face possible rejection. Here, we are reminded once more of Bauman's (1999) suggestion that the future exists like a "target" towards which the individual travels, hoping for fulfilment. If you are single, live alone, and are getting older, the future can also provoke anxiety, which he recognises. Television technology allows Daniel to manipulate time, filling non-work time and keeping him "occupied". He is able to use this to weather the consequences of his decision not to socialise or form a relationship and also meet the demands of time passing as he approaches middle age. This makes the exigencies of the future temporarily disappear, as he is able to exist in a world of the present in which he has some control over how he "passes time". However, this is also coloured by some ambivalence and anxiety, and it is this that accounts for the melancholic tone and feelings of sadness which emerged through the interviews.

* * *

"Twitchy viewing"

Chris (29) lives alone in his own home which is cosy and stylish, with all the trappings and appearance associated with a young professional.

Chris could be described as a cineaste. For Chris, viewing is also a contradictory experience, as leisure time presents a challenge, and viewing often generates anxiety. Unlike Daniel, who clearly delineates his work and leisure time, for Chris the worlds of leisure and work overlap and this intersection presents him with dilemmas. Chris is in the early stages of a career in information services specialising in film. His professional and personal interests overlap. The disciplinary demands of developing a professional identity in contemporary neo-liberal economies influence Chris, as he continually feels that he "should be" working. The thought of watching television is appealing but makes him "feel guilty". As an aficionado, with an intellectual interest in films and also as part of his work identity, he spends a lot of time recording, collecting and cataloguing videos and DVDs of films and television shows, rather than watching them. Chris explains he adopted this practice in response to a time of change involving living alone, worries about his career and the end of a relationship:

> because I was having a bit of a ropey time and I needed something to do. I was kind of *pushing things away* and getting a little bit *obsessed* about stuff. I got really *obsessed* about recording films. I ended up with mountains of videos and all these films I never have a chance to watch. I'm *really anal* about stuff like that you know.

He explains that when he watches television or films at home the experience of viewing alone generates anxiety. He is unable to relax as he is wasting time and always feels he should be working or "doing something productive". Chris explains that when he sits down to watch a film he gets "twitchy, anxious and feeling guilty about not working". He thinks this is something generated in response to being alone:

> Well instead of sitting watching the film I'd be worried I should be doing the washing up or something. And I'd get up and do it half way through the film. And there have been times when ... I haven't even watched the end of the film. I can't relax if I'm on my own. I find that a bit frustrating really because I come home from work and all I want to do is *slob out* and watch a film, not necessarily to totally engage with it – but to *switch off*.

Unlike Daniel, it seems that Chris is not able, or is reluctant, to lose himself in viewing. He cannot allow himself to become fully absorbed and, at times, maintains what might be described as a tangential relationship to the act of concentrated viewing, through his cataloguing and collecting. With his knowledge of cinematic theory, Chris would know that pleasurable film viewing is usually associated with an intense absorbing "gaze", rather than the distracted "glance" of television viewing (Ellis, 1989). However, even this potential absorption seems to be a pressure for Chris as his desire is to "switch off" and "slob out". He even feels anxious about his inability to relax. Partly some of his anxiety relates to the *choice* of text, as he actively seeks out films to watch that frequently generate an unpleasant viewing experience. These films tend to be critically acclaimed "difficult", "intellectual" and cult films. He describes such films as "threatening", because they are something he "doesn't intellectually *understand*". Following a viewing of a difficult film that he does not feel he understands, he then has a new set of demands to "chase it up and read around it and *make myself* understand it". This is unlike the intellectual colleagues he deals with on a daily basis, who he assumes *do* understand difficult films.

Rather than a form of relaxation, his viewing is an important part of the intellectual labour he associates with his professional and personal identity. The interview is peppered with statements about Chris's lack of confidence in his professional abilities. He describes himself as an "anxious" person who worries about his intellectual and practical abilities and being "exposed" at work. He fears that he is a "loser" in his professional life and in relationships. He explained that he is "absolutely petrified" about the future, as his new girlfriend has a very well paid job and he perceives his own economic prospects as poor in comparison to hers. He links this to the potential failure of the relationship saying: "Why would she look at me? She earns three times what I do." It is possible to explore some of these anxieties in relation to class and gender.

Changes in work patterns, familial relationships and social mobility have been theorised in relation to men and what has been deemed the "crises" of masculinity (see Beynon, 2002; Whitehead, 2002; Flood et al., 2007). Men in "crises" are said to be unable to understand or meet the needs of women. Masculinity is marked by "vulnerability and insecurity" as men are constantly forced to question the things that they would have been able to take for granted in the past (Beynon, 2002, p. 135).

Chris is the middle child of three and grew up in a working-class home. His mother was a housewife and his father was a manual labourer and the breadwinner, and this was an important sign of masculinity. The physical strength and endurance associated with skilled manual labour were said to offer working-class men secure forms of masculinity (Beynon, 2002; Willis, 1982) which as we will see below, is something that eludes Chris in his professional and personal life.

Discourses of taste, which often converge around the concept of intellect, have been linked to questions of differential social power (Bourdieu, 1984) and are prominent in work arenas based on intellectual abilities and questions of "good taste". Chris has an older sister and a younger brother. He represents an example of widening participation in higher education in the latter decades of the 20th century. Along with his sister, he is a first generation university graduate, who has entered a profession and achieved what would be deemed as upward social mobility. As the child of working class parents, however, he does not benefit from inculcation into the bourgeois world in which he now operates. One way this translates for Chris is into anxiety generated by the dilemmas posed by relaxation versus intellectual viewing. At the same time, he is constantly trying to deal with difficult feelings about not using his time "productively". Neo-liberal subjectivities are formed in relation to the demands to "perform" and "achieve", particularly for professionals in highly-pressured arenas, but this generates anxiety (see Elliott and Lemert, 2006; Layton, 2003). Chris has not been immune to the demands to perform and be successful, and it is possible to see the pressures he faces as part of the lived experience of masculinity, class and social mobility.

Chris lists a number of films as "favourites" that he acknowledges are, paradoxically, also "threatening" to him. They are recognised for complex storylines, plot, unconventional dialogue and characters whose psychological motivation is difficult to understand. He feels a sense of fulfilment only after he has mastered these films intellectually. One theme, which emerged in relation to a number of these films, is that they contain characters he perceives as "losers". I want to suggest that it is possible to look more closely at specific points of identification with some of these texts to understand the conflicts and anxiety generated by solitary domestic viewing.

Barton Fink (1991, Dir. Joel Coen) is deemed a favourite, but one that creates a disconcerting experience when viewed on TV at home, as it generates anxiety and threat. Chris describes:

> *Barton Fink*, which I really like, but it is *gut wrenching* and *sick making*, because he was such a *fake*, *weak-willed* guy. I find that a bit disturbing really. I *hate* seeing really fake people, really *weak-willed* people who are really unsuccessful and to see that on film it gives me the *shivers* because I would so hate to be in their position – it's almost *scary*.

Given the information that Chris provided in the interview about his fears and worries, the tone of this extract suggests that his fears revolves around points of identification. The film concerns a man's struggles with "intelligence", failure in professional life, relationships with women and feelings of discomfort and aggression towards his working-class neighbour. *Barton Fink* is a moderately successful screenwriter who is contracted to write a screenplay for a large Hollywood studio but develops writer's block. When he eventually finishes the script, it is not well received and he is devastated to be accused of lacking talent. There are numerous scenes in the film where Barton struggles to pass the time and to find a way through his problem. He is often curled up on his bed with a pillow over his head to shut everything out, evoking Chris's desire to "slob out" and not work, when really he "should be".

There are points in the film where the viewer sees close-up shots of Fink feeling *extremely uncomfor*table *as he views films*. The sight of a weak, unsuccessful, fake intellectual man on screen is very difficult for Chris because it presents him with a number of his own fears around exposure of his professional limitations and difficulties with relationships. However, the "gut wrenching" aspect for Chris is not just that he is presented with scenarios that capture his fears; it concerns his *identification* with this flawed character who tries to portray himself as a cultured intellectual. Two of Chris's other favourite films and characters also suggest some uneasiness around his masculine identity, as they involve men who masquerade as women: *Fletch* (1985, Dir. Mike Ritchie) and *Some Like It Hot* (1959, Dir. Billy Wilder). *Fletch* is about a newspaper reporter who writes under the name of a woman and in *Some Like It Hot Daphne*, the character Chris told me he particularly identified with is in fact *Jerry* (Jack Lemmon), a musician, in drag, on the run from the mob, with *Joesephine/ Joe* (Tony Curtis). When I asked him about his identification with Jerry/ Daphne Chris explained that it was because of Jerry's Character. Chris tells me he identifies particularly with *Daphne/Jerry* because he "lets others walk all over him" and Chris wishes he were "better" at standing up for himself, particularly with girlfriends.

Social and physical mobility also change familial relationships, and the processes associated with moving from being working class into the middle class can be painful (Layton, 2003). One distressing feature sees the upwardly mobile individual find aspects of the self and others abhorrent. For example, aspects of working-class taste and disposition. The character *Fink*, who Chris problematically identifies with, finds his working-class neighbour disgusting. This aspect of contemporary upwardly mobile subjectivity can be explored further by examining identification in the light of Chris's biography and by returning to the theme of "threatening" viewing.

We have seen "threatening" viewing for Chris associated with being alone ("watching on my own I find quite challenging"). In contrast, he explains what he calls "harmless" viewing:

> My family would always be about ... if I was sitting watching my dad would probably just be coming home from work and he'd come in when I was watching, for example. And Mum'd be buzzin' about doin' something. ... Television definitely does bring back that sort of memory, that homely safe feeling.

This "safe", distracted viewing is associated with childhood, home, family and the rhythm of the day (Lacey, 1997). Chris also tells a heart-warming story of watching films with his father who would fall asleep, tired after a day's work, describing the intimacy as "bliss":

> My memory is of it just being me and my Dad. And at that age I got on really well with my Dad I idolised him. He'd be asleep next to me. Embarrassing me probably because he'd snore a lot but that was *total bliss* for me. That can be quite emotional thinking about that because I *was* really close to my Dad and I bonded with him.

This indicates that television as an object can serve a "mnemic" function (Bollas, 1992, p. 34), enabling Chris to touch base with previous positive self-states in a way that reduces anxiety. However, his viewing is also entwined in complex processes of loss associated with his social mobility, which is felt most acutely around his painful loss of intimacy with his father and in a particular form of sibling rivalry:

> I've got a younger brother who is just the total opposite to me and he lives about two hundred yards from my parents, and has never been away, never does anythin'. It sounds really harsh on him but, erm, he's married and really conservative and went to work with my dad. He's just really *quite different* to me.

Chris's brother really is his opposite; he is working class, married, unambitious and traditionally masculine. Despite this, he retains a form of intimacy and commonality with his much loved father that "successful" Chris will never experience again in the same way. This familial tension is another element of Chris's ambivalent feelings, phantasies and fears about his masculinity.

Recent years have seen the rise in "credentialism", where middle-class men are located in employment that requires technological computing competency and keyboard skills (previously seen as women's work), now re-defined as a middle-class arena of "competition and power" (Connell, 1995, p. 55). Working in information services means that Chris does not occupy the new spaces of power such as IT or banking where the rewards are high. He occupies a space precariously close to women's work, with low pay and insecurity about his ability to master and control technology, which places him in an uncomfortable position in relation to his new partner. This might also explain why Chris identifies with characters who are men but who might appear as women.

Sennett and Cobb (1972, p. 268) link masculinity to displays of "ability". If a man is educated, he is expected to be able to demonstrate capability and can cope with tasks that demand skill and competency. The disparaged and incompetent man is one character who features in discourses of masculinity in crisis; he performs poorly at things that, traditionally, were assumed men did well such as practical repairs (Beynon, op. cit.). Chris explains he is "useless" at DIY and that his father visits him every couple of weeks to do all his practical jobs. He gave an example of how inept he is with technology by showing me a telephone with lots of complicated functions and explained that, despite working with technology in his job, he "had to ask his Dad" to help him set it up. This contrasts with Chris's brother who works alongside his father every day and is practical. This inability to be competent at practical tasks does serve a purpose, as he is able to use this as a means of maintaining the intimate relationship that Chris had with his father as a child, and it creates a space for them to share time away from his aberrant brother.

I want to explore these tensions in relation to television viewing and return to this important idea that television operates as a transitional object. While viewing can be conflictual and anxiety provoking for Chris, he still manages to find a viewing space and adopts what I will call an "evocative transitional text" which helps to resolve some of these conflicts.

The US comedy, *Frasier* (1993–2004, creators and producers, David Angell, Peter Casey and David Lee), was already well established before Chris "discovered" it and he instantly became hooked. Once discovered, it became subject to the usual treatment: "I started recording *Frasier* every night. I got really obsessed with recording it." This is a text that Chris *can* and *does* watch "over and over again". He uses the text as "wallpaper", explaining, "I have it as background noise or whatever". His use is in keeping with the idea of a sensationally based transitional phenomenon, used to manage separation anxiety, felt most acutely by the child at bedtime (Winnicott, 1971). Significantly, then, this is a text that is repeatedly watched in his bedroom:

> What I do which is a bit sad, is when I go to bed, I've got a video in my room and I'll put on *Frasier*. I'm not necessarily watching it, it's just background noise really, *harmless* [my emphasis], I do that a lot, although I try to phase myself out of doing that. I should go up there and I should *read*, or do something more stimulating, but to be honest, I go to bed to *slob out*.

Clearly, even this soothing text that allows him to slob out still provokes a degree of anxiety about his use of time. Chris is constantly dealing with conflict between "working", intellectual study, doing chores, and the desire to relax and do nothing. He associates self-discipline with dynamism and is constantly disappointed by his indolence. The final sentence of the extract calls to mind the image of a frustrated, "weak" "Barton Fink" lying across his bed struggling with his inability to write. In *Frasier*, however, he has found a text that helps him deal with the contradictions and conflicts between his inner and outer worlds, and its transitional qualities allow him permission to use the text in an anxiety reducing setting.

If we examine character and storyline in *Frasier*, it is easy to see the points of identification but also, I think more significantly, the way in which this text helps resolve contradictions for Chris. The comedy revolves around sibling rivalry with his brother *Niles* and an unconventional domestic arrangement. *Frasier Crane* is a psychiatrist and a cultured intellectual, and despite his supposed superior knowledge about "women", he is unsuccessful with women, having had a series of failed relationships. He is the son of *Martin*, a working-class man who lives with him. While *Frasier* prefers high culture, his father sits in a tatty old recliner chair, drinks beer and watches TV. His father's carer, *Daphne*, is an English working-class woman who, like Chris's own mother, "buzzes" about the apartment. His best friend and work colleague *Ros* is a professional woman but with whom he does not become romantically involved. This friendship has similarities to Chris's description of his relationship with his sister, who he did not get along with as a child. However, as adults sharing the same experience of going to university, this brought them closer together and now they are great friends. In fact, he was staying with his sister when he first discovered *Frasier*, realising it was "really good". The comedy is generated around the themes of class and familial conflicts. *Martin* and *Daphne* frequently undermine

Frasier's posturing with blue-collar values. There are constant clashes emerging out of their intellectual and class differences, which they must negotiate. However, the relationships are held together by a deep bond of love. Despite their differences, there are scenes of love and intimacy between father and his sons. The *Frasier* family lives together happily despite the characters' class differences.

Frasier is Chris's favourite character:

> He's really *calming*. Because he's a buffoon, really. He's obviously really intelligent but he's also a bit of a buffoon. So it's quite heartening to know that even these people who are super intelligent aren't perfect. They can be a bit flaky or whatever.

Frasier is a "buffoon" but Chris finds him calming, identifying with his flakiness, while the text, sending up intelligent people, provides one way of coping with his anxieties about his professional and intellectual identity.

Aspects of the characters' lives, settings, plot development and comedy make *Frasier* an "evocative transitional object". For Bollas (1992) and Miller (2008), object selection articulates something about the subject's self-experience. Objects are outward signs of accumulated life experience (ibid.) Bollas recognises that people can be evocative objects for those with whom they are in an emotionally charged relationship with, but equally the object can be a physical object or an experience like a film which touches the subject on a deep level (Jemstedt, 2011). The distinguishing factor is that the object has been subject to projective processes which colours the experience of the object. Thus, evocative objects, because they come from deep within us, carry a history and bear the marks of parental provision and internal object relations (ibid.). "The decision to use an object rests with the unconscious aims of the person" (Bollas, 1992, p. 36) and crucially, this builds on the importance Winnicott placed on an individual's use of subjective and transitional objects for psychic survival and growth. "Evocative objects" are distinguished by the way in which they contribute to psychic growth and creative living. They are not static as they stimulate "use". Subjects are led to them intuitively, but they are then worked on unconsciously. For this reason, a text cannot be "read" or analysed as a singular bounded object, it is something that is put to use, in this case by Chris, in a specific setting in particular ways, holding potential transformative functions for him. However, this must be qualified because evocative objects can also be "born of conflict" (ibid., p. 79), which suggests something of the difficult situation in which this text was discovered and put to use. The soothing transitional aspects of the text concern the ways in which the "family" members resolve conflicts and continue to love each other and experience intimacy. The humour also helps Chris manage the anxiety associated with his professional life. *Frasier* evokes memories of family viewing as a child to help him manage anxiety as well as coming to terms with issues of loss associated with social mobility. This case study can help us to understand something of the contradictory experience of gender and class and the lived experience of membership of the technological middle class. As I have suggested, this study also shows that viewing is conflictual and points to the differing ways in which audiences can use viewing to unconsciously address emotional concerns and conflicts.

This chapter has outlined some of the ways that television has changed from a shared activity to a more individualised practice, following the advances associated with time-shift and digital technologies. I have suggested that the idea of family viewing is partly a construction and cannot automatically be associated with communication, intimacy and emotional security. However, as the case studies have shown, this idealised version is evoked by these viewers in response to the difficulties associated with being single and living alone. It should not surprise us, then, that for some viewers, like those described here, television should exist both consciously and unconsciously as a site where feelings of loss and desire are evoked, experienced and opened up to the possibility of emotional change. The studies attest to the psychosocial significance of domestic television in the pre-digital era and the psychic possibilities that have emerged in relation to the developments of digital and time-shift technologies. Such developments are also significant when applied to the idea of audiences using the television set in combination with other technologies to construct a gendered, emotionally germane user-flow.

Television viewing is also a socio-cultural activity, and I have described some of the ways that television has been taken up and translated into some common sense, moral, and even dystopian discourses. I broadened the idea of "uses" and "needs" to think about the strike experiences of television as an "object" with specific emotional uses. Also, television was examined as a psychosocial activity along with the relationships with specific texts. I have explored one prominent and specific discourse around the notion of television as a leisure time or "time passing" activity, which in different ways can generate anxiety, as the examples above show. Such studies have the potential to illuminate aspects of increasingly multifarious relationships to work and leisure time in neo-liberal capitalist societies. These accounts illustrate multifarious psychosocial relationships between individual biographies, discourse, social organisation, and internal psychic conflicts and imaginative lives. Psychoanalysis also helps us to think about the emotional consequences of social organisation and change. It also provides methodological insights which help to show some of the ways in which the presence of the researcher has an effect on the data produced.

Watching television to pass the time or relax, then, involves psychosocial processes, and this chapter has used the idea of evocative transitional objects, suggesting that they perform important but contradictory holding and anxiety management functions to illuminate this. Perhaps those much maligned uses and gratifications theorists, with their focus on television as a problem-solving activity, were right to note that viewers use television to engage with emotional conflicts both consciously and unconsciously. Psychoanalysis provides an important set of theories and concepts to supplement sociological approaches that have dominated television study.

4

Favourite Things: Objects in the Life of a *Castaway*

Sophie is often mystified by the fact that her "most favourite" films and programmes cause her to feel sad or make her cry when she views them, even though she has seen them so many times. She is equally perplexed by something she recognises in her viewing practices; taking sick time off work to create a space to view a "favourite", especially when she knows "it will upset me". Sophie explains "I'm not really sure why I put myself through it, you wouldn't really wanna do something that was gonna make you feel sad purposely, but sometimes you do watch things to make you feel sad." This key conundrum is, for Sophie, something she is unable to fully explain or understand.

This idea of viewing as a practice which can be rather puzzling for people is a central theme of this volume. It has been proposed that it is possible to understand why this is the case, in part, by looking for points of biographical identification along with an examination of psychosocial processes that come into play when audiences engage with favourite texts. This includes taking into account history, culture and social processes as well as the form and meaning of the text. Equally, the idea of viewing, where favourites are concerned, as something "pleasurable", has not been excluded, but has been called into question, particularly in its conventional understanding as something enjoyable. This chapter will continue to develop this approach through the case study of Sophie.

Sophie has an attachment to what she calls "stuff" and "things", which for anthropologist Daniel Miller (2008) are expressions of the self and evidence of life events which are inherently "comforting". Miller rejects materialist assumptions that the excess of "stuff" means that people are more superficial and that "things" take the place of relationships with other humans. His central argument is that people who have close relationships with objects also have close relationships with people and this can be gleaned, not through psychological explanations, but through the art of anthropological thick description. Sophie's case study allows us to explore this proposition but challenges the idea that we can learn about the complexity of object use without recourse to psychological accounts. In this chapter, then, I follow Miller, in that I recognise and describe the relational connections and overlaps between the self, other human objects and the "things" that permeate human lives. I take an object relations position, however, and argue for a view of objects as "evocative" in the broad sense that recognises objects as containers for projections, and also that the structure and integrity of the object are important in understanding what aspects of the self are brought into play when objects are intuitively put to use (Bollas, 1992).

Chapter 1 outlined some of the ways objects stimulate subjects, which often relates directly to the structural specificity of the object or what can be thought of as its character or use value, such as the way in which the object kindles or excites the senses of sight, smell and touch. For example, the desire to blow up or hold a balloon might be associated in its use with freedom, flying and fun, with the anticipation and excitement of a bang, which may or may not follow. The idea is that the use of an object will relate not only to the unconscious and intuitive motivation of the user but also to the social meanings and the object's structural integrity which can, as both Miller (2008) and Bollas (1992) acknowledge, also speak to the user. It is only with recourse to the relationship between inner and outer worlds that we can understand the complexity of the relationships with different kinds of objects.

Another aim of this book was to work with an in-depth approach to assess the validity of psychoanalytic concepts and ideas in non-clinical settings, using a method that allows the concept to emerge from the data. Therefore, along with an examination of objects as evocative, the chapter will explore the concepts of linking objects, signal anxiety, mourning texts and repetition compulsion. These notions can help illuminate ways in which some favourite texts can be used in processes of emotional self-survival that are unconsciously designed by the viewer.

Introducing Sophie

Sophie works in administration and has mostly done this type of work since leaving school at sixteen. She is recently married and the couple live in a pleasant two bedroom flat. Sophie had a great deal of instability to contend with throughout her childhood. Her father left the family to have a relationship with another woman when she was aged six. He continued a cycle of returning to and then leaving the family. Sophie's parents have divorced and re-married twice, and Sophie feels her mother has prioritised this relationship over hers. At the time of the interviews, Sophie's parents, much to her disbelief, were preparing to leave her alone again. This time, they were to emigrate to a foreign country. This inevitably coloured the emotional tone of the interviews.

For Sophie, viewing was associated with two forms of stability; emotional and economic. One happy memory from her early life involved the family watching *The Muppet Show* on television together before her father left. Following her father's first departure, the family's economic situation changed overnight. Sophie, her mother and brother ended up sharing one room at her grandparents' home. She was also devastated at the loss of the family's dog which they were unable to keep. In later life this became a point of identification with the character *Dorothy*, in *The Wizard of Oz* (1939, Dir. Victor Fleming), where a narrative theme concerns the potential loss and safety of the dog *Toto*, as I will discuss. The social consequences of her father's abandonment of the family were articulated through a discourse of "normality" in the interviews. This "normality" takes a particular suburban lower-middle class form of an economically secure lifestyle, typical of many at that time, who experienced increasing affluence and the possession of consumer goods as a marker of this. She says:

> I don't know, if my Mum and Dad hadn't separated, if we would be very different. We were very 2.4 children, new cars, holidays and all that and it was different... We weren't rich or anything but kind of middle class, ok, doin' ok for yourself. *We had a pond.* (Interviewee emphasis)

Following the decline in economic security, Sophie experienced other social costs associated with their changing family structure at that time:

> I guess we were a *normal* family when my Dad left. We were living with my Grandma and Grandpa and it was still quite *unusual* at that time. There weren't *that many* people that were not living with their Mum and Dad. I know it's much more common now, but I don't remember it being that common *then.* (Interviewee emphasis)

The family never recovered financially, and when her parents eventually got back together her father's career had taken a downturn and the family could only afford to purchase a "run-down" ex-council house, with all the negative symbolic baggage this carries (Hanley, 2007).

Sophie identifies the 1980s, when she was a teenager, as a really important time for her. In response to her family difficulties she made good friends at the local comprehensive school, and this offered her some stability. Friendships became important at this time and have remained so. Like many teenagers she started to like music and pop groups, television programmes such as *Beverly Hills 90210* (Darren Star, 1990–2000, Fox) and films. It was also the decade that she discovered, via a favourite film, *Back to the Future* (1985, Dir. Robert Zemeckis), "1950s America". Over time this developed into an important fantasy space for her.

Often Sophie "retreated" to her bedroom, which she used as a creative and imaginative space, turning it into a refuge as things were "not particularly good". This was because of:

> Family. Mum and Dad. Dad was there. Then not there. Back again. Then not there again. I was looking for something that was a bit more consistent. You can't really change things that are beyond your control.

Objects were used to help her gain stability (Zittoun, 2006) and the consistency she sought. The combination of popular culture, her bedroom filled with her "posters and music", and her friendships helped her cope. She also found stability through her Grandpa who filled the gap left by her father. He was: "more like my Dad, stable and more there". Unfortunately, he died suddenly and traumatically whilst she was a teenager, and she had never fully recovered from his loss. At the time of the interviews the other half of this precious couple, her Grandma, had also recently died. This, combined with her parents' imminent departure, indicated that the emotional tone of the interviews was influenced by loss, which I could powerfully feel.

The position taken in this volume is that feelings do not "originate" or are "contained within one individual" in fact they circulate and move via projective processes between interviewer and interviewee (Campbell in Redman, 2009, p. 60). Interviewers' responses are therefore furnished by their own identifications and subjective imprint. My own experience of loss allowed me to identify with Sophie's feelings of sadness in the interviews which helped me to understand and empathise. Whilst there was clearly sadness for Sophie, this was not the full picture, as I also experienced excitement and joy when she spoke about some of her "objects", which she frequently uses to try to change her mood. I found

Sophie immensely likeable. She is friendly, chatty and has lots of friends, and it is easy to see why. She has a child-like quality about her with her hair in pigtails and looking remarkably similar to one of her icons: *Dorothy* in *The Wizard of Oz.*

Sophie likes to consume. Her flat is furnished with what I would describe as bright, colourful and "optimistic objects". Her walls are adorned with film posters and Disney prints as she "loves everything Disney". Her home is full of her "stuff", expensive Disney collectable and limited edition items, objects from her holidays to Las Vegas, New York, Florida and San Francisco. The kitchen contains expensive and stylish gadgets alongside numerous mass-manufactured knick-knacks keepsakes or "bits": coffee jars, humorous signs with a 1950s feel, Elvis ornaments and Disney mugs. It is almost impossible to see the fridge as it is covered in decorative magnets. She has a room crammed full of dolls (Disney and themed *Barbie*™ doll "mint in box" limited editions), teddy bears and *Beanie Bears*™. Her bedroom is overflowing with designer label clothes and shoes which are piled high on chairs as she does not have enough wardrobe space. She visits the USA every year to Disney World or New York, sometimes both, to shop for clothes, and the result is usually a large credit card bill which she must then try and pay off. The bedroom is themed in an African style like *The Lion King*, a favourite film that "always" makes Sophie cry and also represents a connection to her grandparents who had lived in Africa. In relation to the point made about the feelings and experience Sophie unconsciously communicated in the interviews and which stayed with me whilst I worked with the data, I remind the reader of the example given in Chapter 2, where I found myself identifying with something that was definitely "not me", her love of these *Beanie Bears™*. I found that, on handling such items, I was able to understand something of the comfort she obtains from their warm and cosy feel and cuddle-ability.

Sophie explains that she often struggles with difficult feelings, and feeling "guilty" for taking sick time off work to watch films and "feel bad". Persistent mood in psychoanalysis is considered to be significant (Bollas, 1992, 1995), and throughout the interviews Sophie would frequently switch from speaking about her "stuff", to films, to her job, how much she hates it and the majority of the people that she works with. This contrasts sharply with her dream job of working creatively as a Disney or Pixar animator, about as far from "admin" as one can get. Frequently returning in the interviews to the elements of this assemblage and the presence of enduring mood suggests these issues are linked. Work is an economic reality for Sophie; it is an unpleasant activity but, as she all too clearly recognises, it is something that must be endured if she is to make her annual shopping/Disney trip to America. So it becomes apparent that there are some contradictory and interwoven processes which Sophie is locked into and which contribute to her mood and shape her activity.

At this point, now that Sophie has been introduced and some context for the discussion has been provided, it is useful to look in more depth at Sophie's favourite texts and how they illuminate her survival strategies in response to the difficulties she faced.

Survival strategies: friendship texts, consumption and the 1980s – *Friends, 90210, Back to the Future, Wizard of Oz*

Sophie explains that all her films and programmes "have a friendship thing" and some of them are "glamorous". Before exploring this further, I want to place the texts that the following sections explore within the context of her teenage years, when Sophie started "buying stuff". This developed as she reached sixteen and economic independence increased with a larger salary on entry into the administration field. Working became something that provided her with money, giving her a new-found form of control over her life. I want to show how Sophie's consumption, which emerged from this, can be seen as a facet of the mechanisms she began to use to provide herself with emotional control and stability. This must be understood psychosocially, and for this reason her favourite texts cannot be "read" conventionally nor divorced from her wider consumption and cultural practices.

To begin then, it is reasonable to presume that a film like *The Wizard of Oz* would become a favourite in childhood. Although Sophie saw it as a child, it did not have any impact on her until she "discovered" it again in her 20s and, significantly, when she was "able to go out and buy it". She has since added to it with a series of expensive "Oz collectables". Similarly, a love of Disney films would also be most likely to initially occur in childhood. For Sophie, however, it is the recent Disney films that she encountered in her twenties such as *The Lion King* (1994, Dir. Roger Allers and Rob Minkoff, Disney) and *Toy Story* (1995, Dir. John Lasseter, Disney/Pixar) and *Toy Story II* (1999, Dir. John Lasseter, Disney/Pixar), that are most powerful for her. Sophie frequently returns to view her favourite texts and, as I have explained, this is often guided by her mood. Evocative objects are distinguished by the fact that they are not static but stimulate "use", and, in this sense, they contribute to psychic growth and creative living often because they are able to contain something of self-states at particular moments of heightened self-awareness. Significant narrative entertainment texts provide Sophie with external representations (used internally also) of her traumatic, anxiety-provoking experiences around attachment, as well as providing representations of the forms of anxiety management and survival strategies Sophie adopted in her teenage years and in early adulthood, and which she continues to draw on.

"Teenage texts", that is texts discovered in teenage years, such as *Back to the Future*, are important, as this is a time when viewing practices that are adopted continue into adult life and capture something of the unique viewing idiom of the emerging adult (see also Daniel in Chapter 3). I would add that in Sophie's case *The Wizard of Oz* is a teenage text, not because it was discovered in her teenage years, but because there is a powerful identification with the teenager *Dorothy*, whose struggles and fears resonate with those of her own teenage and childhood years. This is a film also about "home": losing one's home, going home and the repeated maxim "there's no place like home". This reinforces Sophie's love of her own home filled with "stuff" which has symbolically replaced that which she lost in childhood. In times of loss and upheaval, such as leaving home, gathering "symbolic resources" is an important way identity can be stabilised (Zittoun, 2006) and a form of symbolic order created (Miller, 2008).

Sophie also identifies with specific narrative scenarios which she attempts to use in phantasy to address her own situation. *Friends*, (David Crane and Marta Kauffman, Warner Brothers Television, 1994–2004), *Beverly Hills 90201*, *Back to the Future* and *The Wizard of Oz* confirm some of her strategies, particularly her ability to make *substitute* families or cross generational friendships, to compensate for the failings of her own family. Those texts which are "glamorous" such as *Friends* and *90201* also draw attention to the key contradiction she faces as a working-class woman: not having enough money to be financially secure, as we will see.

Back to the Future and *The Wizard of Oz* are important texts because they take Sophie to "another time and another place" away from "current things" in an attempt to facilitate a change of mood. In the case of the former film, the idea of time travel is appealing and introduced teenager Sophie to the fantasy space of 1950s USA. Sophie explains with a distinctly psychosocial flavour:

> It goes back to 50s America and that's where I would like to be. It just seemed things were more carefree. I want to be a bit more chilled

> out about things and the 50s takes it back. After school, going out dancing, playing a jukebox. It just seems like that's not how I am but how I would like to be, more chilled out. But society and my life doesn't allow that.

Lacey (1999) examined the location of America as a "fantasy space" in post-war discourses. She found that in 1950s Britain, America was experienced as a structure of feeling linked to signifiers of "possibility, affluence, escape and transformation". For many working-class women in the 1950s, the dream of owning material goods and living a more affluent lifestyle, as seen in American movies, was experienced in situations of material deprivation. Sophie has not experienced the 1950s, but she has experienced a change in her economic and emotional security at a time when a growth in materialism and increased affluence for working-class people was accelerated in Britain under the Thatcher governments of the 1980s, a crucial time for Sophie when she was struggling to find ways to cope and survive. If the Thatcher years in Britain are also thought of in terms of a structure of feeling, then it is that it was possible and desirable for all to have an abundance of material goods, and an affluent lifestyle including a stylish home, holidays and cars. These were markers of success and achievement and used to construct and express identity[1]. These are also things that Sophie strives for.

Prior to her father leaving, Sophie's memory/fantasy, was that the family was part of the nouveau-riche middle class; in contrast financial hardship has been a reality that coincided with her experience of rejection and emotional abandonment. This explains the statement that society, which I would interpret as another expression of her economic difficulties, and her life, that is the way in which her past experiences cut into the present, do not allow her to feel "carefree" or happy. This also has a "then" and "now" quality to it where her past loss becomes entangled with her present situation.

Sophie has a relatively secure job that, within the pay scales of her field, is actually well paid for the grade she is at. However, she *hates* it and feels *trapped* and unable to leave. The impact of a flexible labour market, as a feature of neo-liberal economic policies for women, means that it is unlikely that Sophie will be able to give up work, or even go part-time after having children (Beck and Beck-Gernsheim, 2001), if she wants to maintain her "holidays" and other features of an affluent lifestyle. This was spoken about during the interviews as she was considering having a child which meant that she would "never be able to give up work", retrain or change direction.

Sophie dreams/phantasises about a carefree and material security which she associates with 1950s USA, (which was not necessarily based on the facts of life in the 1950s). Tapping into this helps her mood to improve. What is striking about the next extract is that, for Sophie, access to a new and carefree feeling is accessed through material objects:

> If I ever get my 50s jukebox I'm gonna put all my 80s records on it. I could put my juke box on maybe I'd be tempted to dance around with the duster and be a bit more spontaneous. Eventually I would like to have a 50s themed kitchen with a jukebox in it and 50s dining chairs.

Sophie can capture the pleasure and the pain of the 1980s and place it the safe phantasy space of the objects of the1950s. This suggests she is using the objects projectively, and symbolically placing difficult or hopeful bits of the self into the object for safekeeping or containment (Bollas, 1992, p. 34). She is also open to the symbolic ability of the 1950s to key into a range of positive signifiers such as those described by Lacey (op. cit.). Sophie unites her survival objects (records) with this safe and psychically productive phantasy/fantasy space.

Back to the Future, as one of her teenage texts discovered on a visit to the cinema with a best friend, who was there for her when her parents were not, is significant and it can be thought of as a "friendship text". The lead character forms a friendship with an older man, perhaps reminiscent of Sophie's grandpa who stepped in to fill her father's vacant shoes. Of greater significance is the fact that the time-travelling associated with the film is concerned with visiting a time *prior* to one's existence. In Oedipal terms, the film is interesting because it addresses origins; the lead character is involved in a romantic relationship with his mother. More significant in this case, however, is the fact that the lead character's parents' marriage is in trouble and it is his actions and interventions in the past that remedy the problems; this is something Sophie might have yearned to do.

I have proposed that Sophie moves to and fro between past and present experiences where some of her favourite texts are concerned. A feature of the concept of "afterwardsness" is that childhood experiences can be revisited and re-interpreted as traumatic in the light of other experiences and through the lens of adult maturity (Laplanche, 1992, 1999). This concept therefore can usefully be used to help explain the effectiveness of *Back to the Future* as a time-travelling text. Sophie is clear that in 1985, when the film was set, life for her was "not that stable": "I don't think my father was around in '85, I can't remember." This form of defensive amnesia occurs throughout the interviews where Sophie's father is concerned. Whenever she speaks of him, she states frequently that she is not able to remember fully whatever point she was making concerning him. Clearly this is an evocative text in terms of creating a space, in which the child, armed with retrospective knowledge, gains agency in phantasy that was not previously available to them, to intervene and repair a broken Oedipal triangle. It is possible that the child may unwittingly carry guilt in their internal object relations, that they are to blame for the damage to their parents' relationship, and the reparation in phantasy appeases this guilt.

Other favourite texts, *Friends*, *90210* and *The Wizard of Oz* are united by the themes of friendship and surrogate families. In all three texts, the central characters are displaced from secure situations, where identity was relatively stable, to new and difficult situations. Here survival comes from making attachments to alternative family units, that is, friends, friendship groups and extended family (such as grandparents and aunts) which Sophie has experience of. The *Wizard of Oz* is an example of a compelling and poignant children's narrative because it addresses key childhood fears and traumas such as abandonment, separation and loss (Rustin and Rustin, 2001). In keeping with the idea of retrospective recognition of trauma (Laplanche, op. cit.), when Sophie saw the film as a child it seems she could not identify with the traumatic content of the text and she only "vaguely" remembers watching it. It was only on viewing it again, retrospectively as an adult, that the narrative and imagery became evocative in the light of her own experiences of abandonment and her survival techniques and responses.

Here is an example of the work Sophie has done with the text to address childhood distress at "not being old enough to fight for" the family dog when her father left and living with her grandparents during her father's frequent departures. Sophie explains the narrative:

> At the beginning of the film she is running away because they are going to take her dog, they are going to take away that which is most important to her. She's got lots of friends. She's not with her parents in the beginning but she goes and finds other people and they're becoming like a substitute family. She goes and makes friends and they're funny and she's got her family around. That probably reflects what I had to do myself. I have lots of different people that I turn to for different things.

In the case of the *Wizard of Oz*, it is possible to see the evocative and transformative nature of the film as an object put to "work". *Dorothy* is introjected and she is also used projectively: she is courageous and able to form friendship bonds. Thus, these teenage texts confirm survival strategies that began in Sophie's teenage years.

However, it is notable that there is also some ambivalence expressed towards some of the more "glamorous" texts. This is demonstrated, for instance, in Sophie's changed, more cynical tone as she talks about *Friends*: "There's not much family goin' on in that, and like they're out on their own", but unlike Sophie's experience of being out on her own, she says *"they appear to be having nicer lives than the life I've got"* (my emphasis). A similar ambivalence and disparaging tone appears in her voice when speaking about *90201* (a text about wealthy teenagers living in Beverly Hills, California) with some anger directed towards her parents: "That's the kind of life I would have liked to lead. Lots more money, but it's a privileged life living off the back of your parents. My life is completely the opposite my parents have never given me anything." Thus being "given" material "things" is linked to being given love and care which, in Sophie's experience, have been frequently withdrawn and removed. Sophie explains that like *Dorothy* with *Toto*, she "fights for her things" and "stuff" plays a central role in her own very intricate object use, which will be explored in the next section.

Forms of object use: *Toy Story* and *Toy Story II*

This intense connection with "stuff" begins with Sophie's creation, as a teenager, of her bedroom as a haven filled with "things" that were more "reliable" than her parents. Similarly, she feels her flat is also a haven and her home and possessions mean "everything" to her. A considerable amount of time was taken up talking about merchandise which relate to her favourite films and programmes. She has a large collection of porcelain *Wizard of Oz* figures, *Barbie*™ dolls, Disney merchandise (*Toy Story* in particular) and *Beanie Bears™*. In fact, she was once featured in a tabloid newspaper and has appeared in women's weekly magazines, gaining recognition for her collection of merchandise, and she proudly displays the pictures in a frame. Her objects have helped make her special and less ordinary. Whilst I am doubtful about theories of "resistance", it has been noted that within fan cultures, collecting and consumption does operate as an alternative form of cultural capital conferring a certain amount of recognition (Fiske, 1989b, 1992). Although popular culture does not have the same symbolic value as legitimate culture, collectables can also blur the boundaries of cultural capital as they have high economic exchange value (Bourdieu, 1984). Some of the collectables and prints she owns are very expensive to purchase. Disney merchandise, for instance, is of equal importance with the film texts: "I always like the animation side of things but I also like the merchandising side, but especially *Toy Story* because they *were* toys so the merchandising was better than for anything else."

Obviously Sophie is not the only "Disneyphile" (Wasko, 2001) who incorporates merchandise into the home as well as taking part in "Disneyana" (ibid.), the buying and selling of Disney merchandise. Brands, it has been convincingly argued, have been shown to provide "symbolic means of connecting a wide range of commodities to an emotional content, which creates the desire within consumers to purchase those products" (Jordan, 2008, p. 19). In Sophie's case the connection with the Disney brand was forged out of biographical and experiential events, which I will explore further. This creates a less transient relationship than that associated with consumer capitalism where retaining the consumer's brand loyalty is notoriously unstable. For children, Disney narratives have a particular appeal to conscious and unconscious fantasies and imagination. As well as being visually and aurally rich, the narratives frequently confront the child's fears of danger, loss and abandonment as previously mentioned (Rustin and Rustin, op. cit.). Wasko (2001) suggests that the link with Disney is made in childhood and every encounter with a Disney product carries an "aura" that links the adult back to childhood. This explains why Disney retains a special place in childhood memories. This is slightly different for Sophie who, like most children, met the Disney phenomenon in childhood, but her attachment is much more recent and the link back to childhood is not necessarily a nostalgic return to happy times. Sophie's wall might contain a "limited edition reproduction" line drawing of Mickey Mouse in *Steam Boat Willy*, but it is the late 20th century films that have the most resonance for Sophie.

It is evident that events in Sophie's life have shaped her object selection and use which has produced this amalgam of consumption, texts and emotional investment in belongings. In relation to particular film texts, it is interesting to note that the *Toy Story* films contain evocative settings and narrative themes. Both *Toy Story* and *Toy Story II* are set in a child's bedroom in a suburban middle-class American home, reminiscent of the phantasy white picket fence home which Sophie lost when her father left. There is also a narrative theme based around the alternative cultural and economic value associated with merchandise and collectables, as *Woody* (an old cowboy doll) (*Toy Story II*) is accidently put into a garage sale and is bought by a collector and put up for sale to a museum. The *Toy Story* films offer points of biographical identification, as both invoke separation, the importance of friendships and the emotional versus "economic" and cultural "value" of toys as "collectables". This, however, creates some tensions for Sophie which I will discuss shortly. For now, I want to remain with the narratives and discuss a key evocative scene. The pivotal narrative quest in both films is for the toys to return to the safe haven of *Andy's* bedroom; the little boy who owns the toys and also *loves* them. There is, however, a scene in *Toy Story II* which makes Sophie cry and which she finds "very upsetting" whenever she watches. The scene concerns abandonment and also represents a symbolic "discussion" of parents and love.

Woody finds himself in the clutches of a profit hungry collector as the final doll in a set of characters from a television series *The Roundup Gang* which, now complete, can be sold to a museum.[2] A further added layer of significance here is that the doll's value is based on its age and, not surprisingly, the fictional series in which the characters appeared was from the 1950s. *Woody* appeals to the dolls to try and escape with the promise he will take them back to *Andy's* bedroom where they will be played with, unlike the static, yet immortal existence that awaits them in the museum. The dilemma of being looked at and immortalised *or* played with, which is actually the doll's use-value (Bollas, 1992, 1995), suggests something of the nature of the production of what Winnicott (1965a, 1971) has called a "false self". Here spontaneity and creativity, which are necessary for emotional health, are lost in the production of a self that *adapts* to meet the needs of another, in this case the environment of the museum. *Woody* explains that denying their play function or character will ultimately lead to unhappiness. It is better to be played with for as long as they are wanted, than to die emotionally on a museum display shelf. However, *Jessie* tells *Woody* that in her experience his faith in owner/parent *Andy* as someone who will always love him is misplaced, and this leads to the scenario which Sophie identifies with.

Jessie explains that once she was her owner's most treasured possession; she, too, remembered the joy of being played with and loved. One day she accidentally falls under her owner's bed. The girl starts to grow up and *Jessie* is forgotten about and lies un-retrieved for years, heartbroken and forever hoping she will once again be restored and loved. Eventually *Jessie* is pulled from under the bed and she is full of joy, but her ecstasy is momentary and mistaken, as she finds herself in a bag, put out for collection by a charity: *abandoned*. The story is told in flashback montage, which operates in similar ways to the visual features of memory and thereby further evoking biographical memory (Radstone, 2000). The scene is accompanied by the moving song, *When She Loved Me* (Randy Newman),[3] which is in actuality lyrically complex. It captures the nature of a loving inverse parent-child relationship, with *Jessie* ready to perform the emotional functions of a parent to her owner, but likewise with her own childlike needs to be loved, something that chimes with Sophie's object-use of cuddly toys. For example, *Jessie* sings:

> When somebody loved me, everything was beautiful, every hour we spent together, lived within my heart. And when she was sad I was there to dry her tears, and when she was happy, so was I, when she loved me. And when she was lonely I was there to comfort her, and I knew that she loved me.

The song in Sophie's case can be understood as an example of psychoanalysis's valuable insight that the child continues to reside in the adult. The song captures the trauma of abandonment as a sign of loss of love, and the child within the adult needs to hear that she is loved. Jessie sings:

> So the years went by, I stayed the same... And she began to drift away, I was left alone, still I waited for the day, when she'd say "I will always love you."

The song also describes difficult feelings associated with change in the past and the present and Sophie's feelings of responsibility towards her treasured objects. Sophie explains in this extract:

> I cry when she's. under the bed and it's all changing. And the girl is changing all the stuff. Putting posters up and nail varnish. She looks out really expectantly. She thinks she's gonna pull her out and love her again. And what it's done for me is that I'm incredibly sentimental about things. Like that teddy bear [points] sitting up there is [husband's] teddy bear and he's had it since he was a baby. And we had his head professionally sewn back on. And I would *never* move him. Like he's *never* gonna go in the loft. I'd *never* put him away. And I'm really sentimental about things I've had from my childhood. And things that my Grandma has given me. Even hair straighteners she bought me that don't even work. I'm too sentimental with things probably.

> Like I had hundreds of *Beanies*™ in that room. And because of the size of the flat and when we were decorating we just had to get rid of them. And *every single one* of those *Beanies*™ told a story. And I was *gutted* putting them in a bin bag. It was *breaking my heart* (interviewee emphasis). I remember where I bought them, what I was doing, what holiday I was on. And I guess because of *Toy Story* I think they all come alive at night and they're up there[4] chatting about what a cow I am putting them in a bin liner.

The interpretation of the scene by Sophie and the concomitant object use of toys are far from infantile responses. They illustrate the importance of the adult's ability to return to the transitional space and the capacity to use illusion, when selves are confronted by difficulties that they are unable to resolve. The child emerges in adult experience creating emotional dilemmas and, in this extract, we see Sophie reality testing, moving in and out of subjective and objective perception. As the process is on-going and the dilemmas not fully resolved, it is possible to see how this scene as an "object" sits, for Sophie, on the cusp of subjective and objective objects areas. It bears the characteristics of the transitional space and transitional object use as it is associated with toys, which quintessentially sit between fantasy and reality and can be used to allay anxiety. Objective objects are said to be outside the area of projection (Winnicott, 1971, 1990) which serves no purpose for Sophie as she clearly uses her toys and objects as receptacles for a variety of projections around attachment, safety and intimacy. This is in keeping with their use-value as objects which can be cuddled and loved. This is a scene that is evocative because it chimes with Sophie's generalised subjective object use described in the extract. This is because, as Winnicott explains, the subjective object does not have to encounter the limits that a more objective relationship to reality would afford (Winnicott, 1971).

The creative adult or infant does not get stuck in either fantasy or reality but moves between the two areas which are infused with both inner and outer processes. Putting the bears in the loft forces Sophie to confront the historical reality of her abandonment and explains why parting with objects is so painful. She is also forced to acknowledge her adult self that must make choices and take decisions. Usually, she holds on to objects in the sense of an adhesive attachment, as she herself recognises, perhaps this is fuelled by defensive phantasies to avoid confronting the reality of separation. In part, this illustrates the process of the storage of difficult and painful self-states such as abandonment, which are returned to through processes of repetition for the work of change (Bollas, 1987, 1988, 1992). It is these features I will discuss in the next section where the repetitive and binding work of mourning, in response to the trauma of loss, is explored.

Mourning and repetition: *The Lion King* and *Castaway*

"Linking objects" are inanimate objects which are used in mourning and adopted when mourning for an object has failed, or where the subject has an ambivalent relationship with the lost object (Volkan, 1972, p. 215). They can also be linked to repetition as features of melancholia (Freud, 1914) as discussed in Chapter 3. Failure in mourning is relevant for Sophie who has an unreliable relationship with her parents who, although physically alive, have frequently been lost to her emotionally. So, if mourning involved a process of loss, in Sophie's case this is complicated by the fact of her parents' periodically coming and going both physically and emotionally. "Linking objects" can be used in Repetition compulsions to symbolically represent the lost object. When Lacan (2004) used Freud's (1920) account of the child playing the *fort da* game (a game of presence and absence), he concluded that representations can fill the gap caused by absence. Sophie explained, her experience of her father in childhood was of being "there" and "not there" and, in this case, I want to suggest that objects are used to address the painful nature of the emptiness of this gap. Volkan (1972, p. 16) states that object use in mourning usually presents in a person who, like Sophie, has an awareness of an investment in the object but lacks a full appreciation of "what was being symbolized". So we can also see one aspect of her object use as a response to grief and loss in relation to abandonment. Attempts are made to introject the object, her toys, material objects or film characters and texts, which for Sophie had inherently good and potentially "transformative" powers with the aim of reuniting herwith what has been lost. Conversely there is also projection where "the object itself is evidence of the externalisation of what is painful – the work of mourning and persistence of the external object relationship" (Volkan, 1972, p. 16).

Hurvich (1989) focuses on the experience of separation anxiety in early life, emphasising the particular form of anxiety generated by the child's lack of control. Situations which stimulate memories of infantile states of helplessness generate signal anxiety as a form of anxiety whose source is the anticipation of being in a traumatic situation "which is a recognised, remembered and expected situation of helplessness" (Hurvich, 1989, p. 309). Associated fears include fear of the loss of the object and loss of love. Aleatory objects are distinguished by their difference from familiar objects and objects sought out for their experiential value (Grotstein, 2002) and are characterised by the element of surprise. Films, by their very nature, are designed to shock and surprise the viewer, which gives their character as objects an aleatory essence, or what Grotstein (2002, p. 79) calls an "aleatory factor". One feature of this essence is that the object has the ability to surprise the subject with scenarios of biographical significance (see also Chapter 5). For Sophie, this involves film moments as mnemonic objects which generate signal anxiety and their subsequent conversion into repetition scenarios concerned with the trauma of abandonment.

Sophie tells me there is a scene in *The Lion King* that she finds difficult to watch. The scene is "definitely not for children", because it is so "traumatic" and involves a lion cub finding his father's dead body and prodding it to see if it is alive. Narratives and visual scenes such as this, that confront the child with helplessness associated with the loss of a parent, are found in numerous Disney films. Ultimately the child, which might, of course, be a lion cub, a fawn or other animal, faces threat and danger but always survives. Such death or life-threat scenarios will not, in real-life, confront most children. Sophie, however, had the experience of seeing her grandfather collapse and later die, as well as witnessing her father have a heart attack and being unsure if he would survive. This scene, then, like that of *Jessie* under the bed, is associated with helplessness and loss which has a wider resonance in Sophie's life and therefore generates signal anxiety. This leads her to conclude as O'Shaughnessy (1994) would argue, mistakenly, that the film is "not for children", but this conclusion is plausible when placed within the context of her experiences. Interestingly Sophie moved from *The Lion King* to the topic of *The Muppets* and *Muppet* merchandise. She explained that if she were to have a baby she was going to "resurrect" her love of the *Muppet Show* and purchase a "load of new merchandise" to decorate the baby's bedroom. I do not think it is without meaning, if we follow the logic of free association, that she went from a death scene to a discussion of something very much about life. I would also argue that the choice of the word "resurrect" in relation to consumer merchandise says something about Sophie's use of objects to bring the self, and potentially deceased parts of the self, to life.

Repetition compulsion is another germane concept which complements what has been explored already with regard to signal anxiety and linking objects. Repetition compulsions are mental scenarios created to "represent the internalisation of traumas and disappointments experienced by the child" from interactions with "primary attachment figures" and, once created, "are incredibly resistant to change" and for this reason they are lived as "character" (Sophie says: "I'm incredibly attached to things") (Layton, 2004, p. 38). Layton (2002b, p. 7) explains that "to mourn requires a kind of re-contextualising that is both painful and goes against the grain of common sense". I suggest that this is why imaginative film/television scenarios can be used in the work of mourning because they allow the viewer to move beyond rationality, reason and realism. In some theories it is resistance to mourning which produces a repetition compulsion, but I want to suggest that in Sophie's case there is a complicated form of on-going and reactivated mourning, which cannot be fully resolved, hence her repeated work around certain biographically powerful scenes such as those described. "As frustrating as that repetition can be", argues Layton, "therein lies the hope for change, because each repetition is a little different" (Layton, 2002b, p. 7) (see also Layton 2002a). So Sophie's puzzlement at her emotional responses and returning to texts which make her "feel bad" can also be seen as an attempt to bring about change in important affective and unconscious meaning making processes. This can be explored further through the example of *Castaway* (2000, Dir. Robert Zemeckis). I will call this a "mourning text" because it contains examples of signal anxiety, linking objects and is a visual representation of a repetition scenario.

The very name of the film *Castaway* is enigmatic in relation to Sophie's life experiences as it is suggestive of loss and exile, but at the same time *Survival*. This is a film that clearly, like many Hollywood narrative films, contains moments that viewers might potentially find sad. In Sophie's case, however, the words she uses, coupled by the emotional tone of her voice and facial expressions on the telling of her feelings about the film, make this another instance of a text where this particular viewer is reading from the position of her own painful emotional experiences, which are in excess of a preferred or dominant reading from the text alone. The scene in question is described as "horrific" and "traumatic" and she explains that "it always makes me cry *a lot*" (interviewee emphasis). In fact, in the second interview we watched the scene together, and as she had said, she very quickly started to cry and said: "see I told you it *always makes me cry*" (interviewee emphasis).

The narrative concerns Federal-Express employee, *Chuck*. On a work mission his plane crashes and he finds himself the only survivor, stranded on a desert island. He is forced to learn to survive. He finds a number of Fed-Ex parcels from the plane washed up on the beach and they come to play a crucial role in his survival. For example, he opens up the packages and uses the cardboard for shoes and shelter; another parcel contains an ice-skating boot, and *Chuck* uses the blade to make an axe. He finds in his pocket a photograph of his fiancée which he keeps along with one unopened package, as incentives to survive; he will one day see his girlfriend again and deliver the parcel. Amongst these objects he finds a parcel containing a *Wilson* volleyball, which was destined to be a gift for someone. *Chuck* has blood on his hands when he picks up the ball and his handprints make a face on the ball. He begins communicating with the ball, talking to him and using grass to give him hair. The ball becomes a friend/family substitute that helps *Chuck* maintain his sanity.

Eventually, *Chuck* manages to build a raft to leave the island and he sets off with *Wilson* on board. A large wave causes the ball to become dislodged and *Chuck* loses him. At this moment the viewer sees *Chuck's* distress at losing *Wilson* as he frantically tries to hold on to him but is forced to watch him float away. This is the moment that causes Sophie such distress. As we watch together she says, "this is horrific traumatic [ ] Shocking, shocking, poor *Wilson* and it's just a ball." She explains further:

> It's *so* sad 'cause that's his *buddy* and he made it out of a shitty old ball. It's more than just a ball. People can't understand it because it's *more* than just a ball. Because I'm very attached to material objects as well. I'm not materialistic but I guess that's why I've got so much stuff in my house, because I feel there's hardly anything I can get rid of. I just get ridiculously attached to things. It was *everything* to him, his *friend* and his pet. (interviewee emphasis)[5]

This scene is replete with significance around the emotional role played by objects, in particular around the idea of objects remaining as constant and stable and fulfilling the role usually played by human objects. It is clear she identifies with *Chuck* and his use of objects as part of a survival strategy. There are many more layers to her identification with this scene that account for its role initially as an aleatory object and its subsequent role as a repetition and mourning text.

There is a crucial piece of information that illuminates Sophie's response and which she has given me permission to use,[6] her maiden name was Wilson, and it is unlikely that this personal significance would be of no consequence. This adds a further quality to the anxiety generated by the scene as it also brings in the notion that she too might be the object that is lost and not knowing its fate also generates signal anxiety.

Significantly, however, Sophie provides a little coda which stops her feeling upset and she suddenly becomes cheery again as she explains:

> He buys himself another *Wilson* in the end. And he has another volleyball sat next to him.[7] That was his person and he *had* to make himself a friend... I've got a merchandise *Wilson* in the other room. And I tried to make myself one out of a ball.

Thus, thinking about the nature of a repetition scenario, which contains both the trauma, and some sense of the perceived solution however small, it is possible to see this as a scene that for Sophie contains both aspects: her experiences of loss and emotional abandonment, and at the same time the survival strategies which make objects into friends, emotional supports for stability and containers for self-states.

Another comment, which can also be linked to this scene, concerns something Sophie told me about her father, which resonates with the narrative of a man disappearing for many years: "When I was young and Dad went, he went away for what felt like *years before I saw him again*" (interviewee emphasis). Perhaps it had been a childhood fantasy that her father was trapped on a desert island and would one day return and it was this, rather than his lack of love for Sophie, that prevented his return. This idea, that the parent has not abandoned the child and that there is something preventing the parent from being with the child is common in children's literature (Rustin and Rustin, 2001).

Thus Sophie's trauma has emotional, economic and social features. She encountered loss at a young age against the backdrop of a social world where consumption and the possession of "things" were linked to identity, stability, fulfilment and security. Clearly these are not separate features: economic hardship is emotionally experienced in such a way that it has traumatising effects for Sophie. For Sophie, however, it has taken the form of a particularly exaggerated form of consumerism. Consumerism has been understood as a form of misrecognition (Zizek, 1989) or commodity fetishism which makes people "defensively substitute the desire for *things* (my emphasis) for the desire for love, recognition and feelings of worth" (Layton, 2004, p. 38). This is manipulated by marketing strategies such as branding, which mask social relations and attribute *transformative* power to objects or things, which is part of the success of modern consumer capitalism, creating connections between individuals' emotional agendas and particular forms of consumerism.

In health we can be open to the use-function and value of an object, which includes projection into the object; continual projection, however, depletes the self (Bollas, 1995). Where trauma has occurred, the subject continually projects wants and needs into objects, and the object can become the only perceived source of having needs met. In Bion's terms the hope is for objects to "contain" and do the difficult emotional work of the mother, a human object, particularly around nullifying fears. This is the potential of a transformative object; it is able to initiate change when the child cannot do it for itself. This process is doomed to failure, however, because texts or other inanimate objects cannot do the emotional work of the primary relationships, except in a limited way in phantasy. This scene addresses Sophie's responses and the solutions she has adopted. Here the tears come in part from a realisation of the fragility of this strategy of using consumer goods in response to the trauma of abandonment. This work of repetition ultimately becomes conservative object use because there is confusion where solution is concerned.

Sophie explains that the story has a happy ending as he "buys himself another Wilson in the end". On one hand this confirms that she is always able to replace much-loved objects with others. This could also constitute the basis for an ideological and critical reading of *Castaway* as an endorsement of American consumer capitalism. This is supported by evidence seen in Sophie's sudden mood change, from grief to light-hearted optimism, when she tells me that when *Chuck* returns home, he makes a fresh start by delivering the unopened parcel. He reinvests in his role as a company man, and drives away with a brand new *Wilson* ball that he has purchased; as Sophie is well aware, all mass produced objects are replaceable. Sophie's mood also lifted as she explained how she attempted to make a *Wilson*, out of a ball, for herself. This act of creativity, which is also a demonstration of her artistic ability (the "only thing I was good at" when she was at school) enables Sophie to convince herself that she is not subject to the criticism that she is "materialistic" because she does not recognise that consumption is crucial to her emotional survival, she experiences it as an expression of creativity. As long as she can purchase *objects*, she is alive emotionally and in control in a way that was impossible for her as a child. This ultimately embeds her in processes that link her emotional survival to consumerism even if she does not perceive it in those terms.

Sophie's study also illuminates some of the emotional difficulties associated with the economic realities of capitalism, such as unemployment, which economically preclude Sophie from making any changes to her much-hated job and employment status. It pays particular attention to consumption and object use as a feature of individualisation which demands one has "control of one's own money, time, living space and body" (Beck, 1992, p. 92). In this case "control" is an emotional phenomenon which is channelled and serves the demands of consumer capitalism. It also explains psychosocially the ability of capitalism to present consumer goods as evocative objects with the power to transform, meet needs and to perform emotional work. This shows that the idea of consumers as ideologically interpellated by commodity fetishism, holding out the promise of the satisfaction of unbridled desire, is a rather narrow perspective, as it is unable to fully account for the internal processes that are multifarious and unique to each individual. Thus in Sophie's case consumer goods are being used in part to supply solutions to personal dilemmas and traumas. This also contributes to our understanding of consumerism and the power of marketing strategies in contemporary capitalism. Ultimately, objects cannot fulfil their promise, but the hope that they hold out means that there is always another object one can move on to – keeping the wheels of consumption turning.

Conversely, in some approaches to audiences within Cultural Studies, Sophie might be conceived as an active and creative consumer resisting the effects of power and ideology in a process of semiotic creativity (see Fiske, 1989a, 1989b, 1992). Clearly this description would be appropriate to Sophie, who is extremely creative and side-steps work in small-scale tactical assaults on power such as using sick time to create time and space that is her own (De Certeau, 1984). The overwhelming emotional responses to her father's continual abandonment of the family, and what she sees as parental selfishness, have resulted in Sophie's strong and painful attachments to objects and consumption practices such as collecting soft toys, merchandise and collectables. However, these processes are contradictory as they lock Sophie into economically frustrating patterns of feeling trapped into an utterly creative-less job she loathes, credit card bills and all the other oppressive veracities of life in contemporary consumer capitalism. Her object use can also be connected to her perception of normality as linked to the possession of material goods and an affluent lifestyle. In turn this must be placed in the context of her experiences of growing up in 1980s Britain under the New Right, Neo-Liberal Thatcher governments whose cultural impact were significant (see Hall 1988; Hall and Jacques, 1983). In this sense, then, the term resistance is given a critical realist psychosocial perspective.

In this chapter the psychosocial approach used has explored viewing in relation to historically significant relations and symbolic signification within the 1980s in particular. This examination of Sophie's favourite texts has shown how psychoanalytic concepts can be used to enhance our understanding of identification, where favourites are replete with biographical significance and both conscious and unconscious meaning-making. The study has added to existing ideas about the way in which texts are "read", noting in particular the importance of emotional and biographically significant readings. Entertainment texts provide visually and narratively rich source material for psychic engagement and can be used by viewers to address emotional difficulty and trauma. Viewing and favourite programmes and films can also be located within wider patterns of object use and consumption.

5

Mothers, Sons, Siblings and the Imaginative World of Working-Class Women's Viewing

So far, we have explored some of the characteristics of favourites such as revisiting texts and repeated viewing. This chapter features two women who have both had powerful reactions and moments of surprise towards something they were watching on TV at home. This element of surprise has been linked in previous chapters to the idea that films and programmes possess an "aleatory factor" (Grotstein, 2002, p. 79) as part of their character as objects. For Bollas (1992, p. 26) the everyday is constructed around those objects and experiences we actively seek out with conscious and unconscious motivation, but also by the aleatory movement of objects. For our purposes in this chapter I am going to follow this and propose that texts, as aleatory objects, are objects that unexpectedly compel the subject to experience something that resides in the content of the text/object, whilst evoking from within the "unknown depths" of the self, something about which the viewer might have been hitherto "unaware" (Grotstein, 2002, p. 79).

The chapter will explore Sue and Mary's aleatory experiences and examine these in the light of the relationship between significant biographical events and identification. One important theme to emerge from the data featured in these case studies concerned the impact of the birth of siblings. Psychoanalytic film theory had attended previously to developmental moments such as the Oedipus complex, which prioritised the parental couple, and its role in gender identification. In this chapter I am going to develop this and explore the impact of the birth of siblings as a key developmental moment, providing the child with an awareness of parental sexuality. At the same time that the birth of a sibling irrevocably changes the external family dynamics, it also has an effect on internal object relations. Therefore, I will explore the way in which family viewing, as something involving unconscious internal relations, imaginatively imbues viewing.

Sue and Mary are both middle-aged, working-class mothers of sons who are getting ready to leave home, and this change requires a renegotiation of their identities. The chapter will explore textual moments as containers for a range of projections and intergenerational identifications that are generated in response to identity in transition. The interview situation provides Sue and Mary with an opportunity to explore these moments both consciously and unconsciously. Whilst each case is unique, providing idiosyncratic findings, important conceptual themes emerge which connect these cases together and proffer ways of understanding identification, so for this reason the cases are explored alongside each other.

Introducing Sue and Mary

Sue is lively, animated and fashionable. She has three children Jack (19), Faith (17) and Henry (10). The family is working class. Sue has spent all her life in the area in which she grew up and the family live in their privately owned home. Sue had a "happy childhood" but the family were "not well off". Her parents live close by and she has one brother, 12 years her younger, who is unmarried and settled in London after graduating from university. Like "most of her friends" Sue, "married young and had babies". When her daughter was beginning junior school Sue thought, "I'm not ready for her to go", and she "suddenly wanted another baby" resulting in the birth of her second son.

Having a child after a considerable gap was something Sue had previous experience of, as she had been an only child for almost 12 years, when her mother announced she was pregnant with Paul, her (only) brother. Sue found this incredibly traumatic at the time and was *"absolutely devastated"* at the news. She remembers putting her head on the table when confronted with the news and "just sobbing and sobbing". Once he was born, Sue had to face a degree of displacement, as "this baby just got all the attention" from her mother.

Sue described herself as a child who was "no trouble" to her Mum but her decisions to leave school, get a job and then marry were not popular with her parents. Her mother allowed her to leave school only on the promise she would continue studying at night school, which she did, and she passed an A level. Once her children were older, Sue started an Open University degree, which remains unfinished. Sue expresses regret about not having continued in further education. Sue is a fan of

Christopher Eccleston. He is an object of desire she "keeps secret" from her family and friends.

Mary has been married to Phil for "twenty five happy years". They have one son Sam who recently graduated from university and is engaged to be married. They live in a small cosy terraced house. In material terms they are not wealthy but, socially, the family are well known and well-liked. They have "lots and lots of friends", and this provides a sense of their importance and success.

As Mary was growing up, she says she was described as "loud" and "confident", but she spent her teenage years feeling "lonely in the family". These feelings of being an outsider began when she lost her position as an only child at age eight. Her first sister was born, closely followed by another sister a year later. This change was traumatic for Mary as following this she felt "very much one on my own out of five". As a teenager she was put under, what she felt was, "immense pressure" from her parents to "deliver educationally". Her father "put the most pressure" on Mary "more than anyone has ever put on me in my entire life". There were rigid study times when Mary would be forced to study in her room for three hours after school. Ultimately she still feels "vulnerable" around him, but also has always had an "intense bond" with him. "There was always this undercurrent in the family" that she should "marry a doctor or lawyer and live in a big house". So she disappointed her parents as she married a manual labourer and did not make the grades for university where she had wanted to study Psychology. This left her with a "fascination for psychological films" and programmes.

There were some benefits to the large age gap for Mary. Her father worked and was often absent. Thus, as the eldest child, she "quite often got to watch things with my Mum that ordinarily I wouldn't have been able to watch". This was remembered as a welcome relief from periods of enforced study. Her relationship with her mother was not always easy and she was not allowed to formulate her own opinions. A particular favourite was the drama set in a female prisoner of war camp; *Tenko* (BBC, 1981–1984). Mary's favourite character was a woman named *Bea* who was seen as a strong character and a "bad" girl on the surface, but insecure and sensitive underneath. Mary saw herself in her teenage years in much the same way: "confident" and "brash" on the surface, but underneath sensitive and lonely. "Layers" is the name Mary gives to this feature of her tastes which involves probing and looking below the surface, moving beyond first impressions. She is drawn to programmes and films which allow her to explore "good" and "bad" girls. Julia Robert's portrayal of a prostitute, *Vivian*, in *Pretty Woman*

(1990, Dir. Garry Marshall) standing out as her favourite film. As a teenager watching TV at home, her identification with "bad" girls always had to be hidden from her judgemental mother who "wouldn't have approved of Bea" as a role model. For example:

> We had conversations [about *Tenko*] and I agreed with every word she said because that *wasn't* the done thing to say *you'd* seen something else whatever, or *you* thought she [Bea] was something *different*. I thought there was more to her than what you could see on the surface.

Mary is aware that she also played an important role for her mother but one that she was not entirely comfortable with:

> I was kind of a bit of a confidante to my Mum quite a lot of the time. I realise now growing up she'd told me quite a lot of things that really weren't for *young children* to hear to be honest. I think I was her company when my Dad wasn't there. She would take me into her confidence she'd chat to me about the things that she really should've talked to her friend about. By the time I was fourteen I knew a lot of things probably fourteen year olds shouldn't know or should *ever* know really.

Mary doesn't say what these things are, but it is clear that she was privy to information she was not mature enough to manage.

Mary remembers having a "vivid imagination" as a child and being prone to "sleep-walking", "nightmares" and "crying a lot". Television programmes such as *The Singing Ringing Tree*, often featured in her dreams. Mary married an easy-going man who is "happy to let her take charge". She feels her relationship with her only son Sam is very different to the relationship she had with her parents. Mary and Phil are extremely proud of Sam, and she explains passionately that they *"just adore"* him.

Both Sue and Mary have often been unexpectedly confronted emotionally by films they have watched at home.

"Seven or Eight" and "Eleven or Twelve": siblings and sexuality

This sub-heading refers to recurring phrases and narrative themes within Sue and Mary's interviews. These themes emerged at different points across the interviews and I have analysed the statements in keeping with the FANI method, in relation to the whole rather than as logically and chronologically ordered statements. It is interesting to explore why these themes should emerge from a request to talk about favourite films or programmes and why they are associated with viewing memories rather than particular texts that are deemed favourites.

Mary describes some "upsetting" and "frightening" scenarios relating to film and television when she was "seven or eight". Sue tells a number of stories of herself and her brother at "eleven or twelve". In Sue's case, 12 appears to have some biographical significance. It was around that age that her brother arrived, she had another child when her middle child was nearly 12, and her favourite film, *Big*, involves a 12-year-old boy. This suggests that the significance of events occurring when Sue was 12 extend into her adult subjectivity. Biographically, the age of "eleven or twelve" for Sue and "seven or eight" for Mary is associated with their change in status, instigated by the birth of a sibling, from what in the 1960s and 1970s would have been an unusually long period of being an only child. This chapter will explore the significance of siblings and how they shape identification and meaning for these viewers. Also the data provides some interesting perspectives on the kind of research that has concentrated on cause-effect and detrimental effects of children watching films aimed at an adult audience. This section presents some interesting retrospective perspectives on this, as both women present stories of being terrified by adult texts as children, and it is my contention that a biographical perspective can help us to understand the unique and personalised nature of this fear.

Sue: *Hush, Hush Sweet Charlotte* (1964, Dir. Robert Aldrich)[1]

On a visit to her grandmother's home around the time of her brother's birth, Sue had a frightening and unforgettable experience. She was allowed to stay up and she saw a film which: "Had such an impact on me. It scared me to death. Do you know I used to go to bed shaking... oh for weeks and weeks?" Sue gives this account of the narrative and frightening moments:

> Bette Davis played this character, she was going to get married and for some reason her fiancée was murdered by having his head chopped off and his hand or something. I can remember the scene with his

> head rolling [with actions] down the stairs. I can't remember how the film ended. I think it was all in her mind. I think it was the *sister* [my emphasis] trying to drive her insane. It was all set up by the *sister* and this bloke set it up to make her think she was insane. I didn't know that until I saw the end by which time I was absolutely *terrified*.

Rosenbaum (1998) argues that out of the innumerable early life experiences a small number of powerful childhood memories, such as the above, are retained. These memories play a "pivotal role" in the life story. The importance is found in both what is revealed, but at the same time, what is hidden; this can be thought of in terms of manifest and latent meaning. The accuracy or the facts of childhood memories such as this are less important than the *affect* and emotional significance of the event (1998, p. 69). Indeed, some factual inaccuracies, when viewed in relation to other evidence present in the data, can be seen less as inaccuracies and more productively as forms of parapraxis. For instance, Sue remembers that it was Charlotte's *sister* that plotted to drive her insane. However, Charlotte was, like Sue, an *only* child. It was Charlotte's cousin who plotted, to force her into suicide to inherit her fortune. If we place this in the context of Sue's fears about the impending or recent arrival of a sibling after almost 12 years as an only child, it is possible to see this as an example of the way internal unconscious phantasy, as part of a response to significant life changing events, permeates the interpretation of this text.

Recently there has been a reassessment of the central role played by parent-child relationship in psychoanalytic thought which means that lateral relationships with siblings and also peers have been neglected (Mitchell, 2003). Chapter 3 introduced the idea of sibling rivalry, showing how relationships with parents and siblings are part of on-going psychosocial negotiations around identity change and loss. The classic view developed by Freud states that the sibling relationship is characterised by envy, hostility and conflict, because siblings are always in competition for a parent's attention and threaten to replace the existing child in the parent's affections. The presence of an actual sibling or even, for an only child, the potential threat of one to come causes hostility and is traumatic for the existing child, not only because of the fear of displacement, but because "it indicates that the mother has had a sexual relationship with a father" (Mitchell, 2003, p. 10). The sibling "is *par excellence* someone who threatens the subject's uniqueness" (ibid., p. 11). Even when a sibling is loved, in the eldest child there is always an accompanying hate as the child realises it can never be the only baby again. Second children realise they will never hold the privileged position of the first-born child. Therefore, if siblings can provoke intense hatred, they *must* be significant to the inner world, and as Coles (2003) has argued, we have not paid enough attention to the way the texture of the inner world is shaped by siblings. In Sue's and Mary's cases this loss of uniqueness is intensified due to the large age gap between them and their siblings.

In contrast to this view of sibling hostility, theorists have argued that relationships with siblings are not secondary to parent-child relationships but have their own intrinsic value. Siblings are introjected and form an important role in the development of self and identity (Hadfield et al., 2006). Identification with siblings plays an important role in psychic development (Coles, 2003, p. 85). They raise crucial questions about similarity and difference. There are never just separate children, siblings are always one of a set and inextricably linked; the same, yet different, pointing to the tensions in identity formation between belonging and differentiation in families. Sibling relationships exist as part of internal object relations and are subject to unconscious phantasy which, in turn, influence the way in which class and gender are experienced and become meaningful. I want to explore this further and show how sibling relationships can contain a mixture of hostility, love and passion through an examination of viewing memories.

Kleinian theory points to the aggressive nature of unconscious phantasy in children particularly in response to the threat posed by a sibling. In *Hush, Hush Sweet Charlotte* Sue imagined that *Charlotte's* emotional and physical existence was threatened, like her own, by the arrival of a wicked sibling who could potentially annihilate her. In response to this threat, she explains that she retaliates and "imagines herself" into the film and says: "I *was Charlotte* and I murdered that man." As we have seen, in phantasy there is no necessary correlation between gender and on-screen identification. The spectator can move around the scenario and occupy multiple positions of identification with characters, plot and even inanimate objects (Cowie, 1997). Sue identified with *Charlotte* because, in phantasy, Sue perceived she had brutally murdered her male sibling in response to the threat posed by his arrival. At the same time she is likely to have been frightened by her own powerful primitive phantasies of death and destruction towards her sibling, parents or both.

Big (1988, Dir. Penny Marshall), *The Birds* (1963, Dir. Alfred Hitchcock) & *The Singing Ringing Tree* (1958, Dir. Francesco Sefani, DEFA)[2]

It has been argued that cinema spectators, (I would substitute the word viewer here), carry a "cinema/viewing" history which is returned to and re-translated in the light of biographical experiences and events (Sutton, 2004). When spectators return to a text, their history is expressed in both diachronic and synchronic terms, of the moment of the event and of the view of the memory from the current position as a form of après-coup or afterwardsness (Laplanche, 1999; Sutton, 2004). Commonly, the term is used to explain the nature of sexual trauma, which I suggest can be associated with sexual awareness associated with sibling birth. Trauma is said by Laplanche (op. cit.) to have two moments. First the actual event, such as witnessing the primal scene, which is not perceived as traumatic at the time because the child lacks the developmental understanding to realise the sexual nature or consequence of the scene. It is only at a *later*, more developed stage that the child / adult realises that what he saw was traumatic. It is possible to see examples from Sue and Mary's viewing histories which show attempts at the re-translation of events as Sutton (op. cit.) has proposed: some, which recognise a degree of the trauma of sexual awareness, as we will see in Sue's texts discussed below, and others (Mary's texts) where a screen memory prevents the full implication of an event being recognised. The difference I would suggest lies in the age differences with Sue being more mature when her sibling was born than Mary.

Big is Sue's favourite film because it makes her "laugh" and at the same time she "always cries". The film revolves around *Josh*, a 12-year-old boy, who makes a wish on a fairground machine to be "big" and wakes the next day to find he is a 30-year-old man whom his mother then mistakes for her son's child abductor. *Josh* manages to convince his school friend of his true identity as a 12 year old boy, despite appearing to be an adult man. With his friend's help, he flees to New York City until the problem can be resolved.

By chance, *Josh* obtains a job in a toy company where he quickly rises to the top because of his child-centred thinking on toys and play. Much of the humour revolves around the misunderstandings between the adults and a child who looks like an adult man. In particular, he becomes romantically involved with a female executive, whilst lacking the maturity needed for an adult relationship, making this a text that is particularly evocative of the processes of translation associated with après-coup. Eventually, he finds the magic machine again and turns back into a boy which, for Sue, is a powerfully affective moment which I will discuss later.

I have noted that the film revolves around the evocative age of 12. There is one scene she especially finds funny involving *Josh* and the woman who he is about to enter into a relationship with. Here the humour revolves around the different adult and child perceptions of events:

> I mean he's only a little lad really at the end of the day, and she's a sophisticated woman and she says, "where exactly are we going" "Where's this relationship going?" Well he's only a young lad. And then it clicks, what she means, and it means "I fancy you." Oh you must remember this, where he starts beating her up with a magazine. Well you know what it's like with teenagers, well kids, if they like each other they lark about and start beating each other up and that's what he does, he starts bashing her with a magazine and *she thinks* that's what turns him on!

This story was remarkably similar in content and humorous emotional tone to a story Sue had told me earlier in the interview. She described a "risky" situation in childhood, where humour diffused a difficult situation. Sue witnessed her father hit her Mum on her bottom during an argument. Sue explained she "had to leave the room" not because it upset her but because it was "so funny". Both stories contain hitting with an underlying sexual connotation. In both accounts, there is a growing awareness of adult sexuality from a child's perspective and in both stories the humour makes the moment safe. Laplanche and Pontalis (1973, p. 335) explain that the primal scene is particularly traumatic for the infant. It can be witnessed, where the child sees the parents in the act of intercourse. It can also be *inferred* "on the basis of certain indications and phantasies"; most significantly it is interpreted by the child as "an act of violence on the part of the father". The two stories are therefore linked by inferred sexual relations through physical violence. Sue identifies further with the awakening knowledge of sexuality that occurred for her at a similar age.

The theme of sexuality and trauma continues in Mary's data as she remembered a frightening experience when she was "about seven or eight" concerning Hitchcock's *The Birds*:

> I remember coming downstairs in the middle of the night.[3] I don't know whether I was upset and I couldn't sleep and I walked in on

> my Mum and Dad who were watching Alfred Hitchcock's *The Birds. And it was the bit where the lady sat at the bottom of her bed and had her eyes pecked out* [my emphasis]. And that gave me nightmares for years and years. I have to say by then the damage was done. It was a film they were enjoying and they were quite engrossed in. And I wandered in and I don't quite know why I wandered in. I used to do a lot of wandering when I was little. I did a lot of sleepwalking but I was definitely awake at this point. I'm guessing I was *nearly eight* [my emphasis]. And I walked into the room and in the time it took for them to realise I was in the room and obviously there was no remote control, you'd have had to jump out of your seat and run over and turn it off. And in the time that any of that may or may not have happened, I'd been turned round and ushered out of the room, I'd walked right in at the bit where the lady is sitting on the floor at the end of the bed and the birds are pecking her eyes out and that was enough. I had nightmares for a long time after that. I can remember asking my Mum about it, asking her why the birds were attacking the lady.

Mary then said that her mother's response to her questions was inadequate (this will be discussed below). She explained her experience of this film as: "children feeling the vibe of emotion but not understanding what is going on". This shares some characteristics with what Freud called the "screen memory" (Freud, 1899, 1901), which is a memory serving as a defensive function because it hides some potentially threatening repressed unconscious content. The memory might conceal a contemporaneous traumatic event or stand in for a memory of a childhood event (Smith, 2000). Rosenbaum (1998) argues that many childhood memories are in fact screen memories, with content that seems factually trivial and unimportant, but are distinguished by a strong emotional feeling that the adult does not understand. Freud (op. cit.) explained this as a process of displacement where significant impressions and traumatic events have been repressed.

Mary was not reading the film's narrative but responding to a visual moment. If we examine the language and tone of the memory we can observe some of the characteristics of a screen memory against the primal scene and the awareness of parental sexuality. Mary walked in on her parents who were "engrossed" in something. It took them a few moments to realise she was there and for them to then react by quickly jumping up and stopping what they were doing. This is strikingly similar to what would happen had the child walked in on its parents having sex. In Mary's case this is reinforced by the fact that she says she was "seven or eight", which would mean that one sister had already been born and due to the very short age gap the next had probably been conceived. Curiosity about origins would have been significant for Mary at this time. She also notes that she had often been sleepwalking and "done some strange wandering about at night" when she was a child. It is quite possible that there had been other occasions where she had become aware of her parents' sexuality, perhaps walking in on them or hearing something, and the film moment becomes a screen memory for this. The trauma is heightened by the momentous consequence of the birth of her siblings who displaced her, a separation she did not fully come to terms with.

Within the Kleinian tradition it has been argued that children have instinctual knowledge of sexuality and of a penis and vagina meeting without the "actual witnessing of the primal scene" (Hinshelwood, 1989, p. 297). This knowledge is akin to a form of primitive affective and embodied unthought known, as the child is thought to have the capacity to understand the benevolent and malevolent aspects of the relationships between parental objects even when the infant is linguistically and developmentally immature. This, I think, is captured in Mary's statement that she *sensed* something without being able to fully comprehend or articulate its significance. Undoubtedly, the meaning of this event, to whatever degree for Mary, has to be considered as emerging from a mutually constitutive process between text and child viewer, and the content of the text and what Mary saw on the television screen is significant. For Bollas (1992) the emergence of the unthought-known is also linked to the aleatory factor of object encounters, and we can link this to some of these frightening childhood experiences as well as those Mary encountered in adulthood (see below). However, there are also resistances to these encounters because of their potential to bring about "catastrophic changes" through the difficult thoughts they allow, in Bion's terms to "be thought" (Grotstein, 2002; Grinberg et al., 1993).

Voyeurism in the cinema is theorised as a form of pleasurable looking which allows the spectator to engage safely in forms of prohibited looking. If this is a form of screen memory for Mary, then the image of the woman with her "pecked out" eyes signifies that she had seen (in reality or phantasy) something that was forbidden. Of course in this scene the consequences being that the viewer/Mary is punished. The quest for proscribed knowledge of origins and the punishment for curiosity are central features of the Garden of Eden, Tower of Babel and Oedipus myths, with blindness as the specific punishment in the latter of these (Grinberg et al., 1993). The scene allows a degree of displacement to occur, with her shock and fear revolving around the screen images rather than her witnessing or phantasising a primal scene.

It has been theorised that memories of the past have a visual component; when we think about the past we *see it* and we feel it (Scott, 1996). As a powerful visual medium, films provide a rich source of sensory materials which can be worked into meaningful experience. Sensations, what we see, hear, smell and feel, are not meaningful on their own: it is only when they are linked up or enabled by what Bion (1962) calls Alpha-elements that the elements become meaningful and can be used in forms of "thought". Alpha-function and Alpha-elements explain how this raw sense data is turned into components that provide the psyche with material for dreams and other unconscious work as well as conscious thoughts (Bion, 1962). It is the coming together of sensory data elements that allows thinking to occur, and this model differs from one that holds that thoughts can occur just because the mental apparatus for thinking exists. This scene from *The Birds* had a sensory linking function, which functioned like a jigsaw piece, making experience both meaningful and frightening.

Alpha-elements may cohere or separate at the "contact barrier" (Bion, 1962), a kind of semi-permeable membrane that keeps conscious and unconscious, sleep, dream and wakefulness separate, and allows the subject to have a sense of the past, present and future (Grinberg et al., 1993). The contact barrier is different to the notion of the pre-conscious, which Freud mainly used to describe a kind of gate-keeping function for repressed contents of the unconscious. The contact barrier *is* protective in that it maintains a balance between the exchange of stimuli and material between conscious and unconscious, preventing too much distortion of reality from phantasy for instance – but at the same time allowing enough movement for thought to be formed to help the subject function in the social world, as well as internally. It can be argued that the sensory richness and intensity mean television and film operate close to the contact barrier and are particularly useful for thinking and thought generation, even when this is unconscious, which chimes with the notion of the affective cusp formation associated with television in Chapter 1.

In response to her fear and to help her make sense of the film, Mary sought reassurance from her mother. For Bion (1962) the mother functions as a container for intolerable feeling, fears and frustrations projected into her by the infant. Significantly, she not only contains them for the child but mollifies them and returns them back to the infant in a manageable form. As Mary explained, her mother was unable to contain or assuage her fears on this and also other occasions. In fact, her mother often explained things in a "very adult way" and "in quite a lot of detail". She tried to "rationally explain" this film to her. Mary remembers that even though she was very young "it would've been very much taught to me as a *grown up* (my emphasis) even at that age". The young Mary coped with this thusly: "what I did a lot of was to actually make up the rest of the story myself because I was quite shocked". Mary was given an explanation she was not mature enough to understand, and presumably this connects also to Mary's uncomfortable feelings about her mother using her as a confidante and friend as she became a teenager. Her mother, for whatever reason, was not able to adequately contain Mary's fears which remain stored in the film moment. Films/texts are of course different objects to human objects as the latter are involved in mobile processes of unconscious communication. The film cannot know what the viewer feels. Whilst films/texts can be projected into by viewers, the film cannot do the work of transforming and assuaging phantasies, frustrations or fears. They can only operate defensively, storing something to reduce anxiety or for protection to, perhaps, be revisited at a later time in response to further experiences or mood; texts are, therefore, containers which cannot contain. What is interesting, and is a mark of Mary's resilience, is that left with no option, she made the story and the particular moment safe by inserting it into a self-created narrative structure she could cope with.

The Singing Ringing Tree is another text which she associates with the time she was "often distressed" and doing "extraordinary things" such as sleepwalking. This can be seen as another example of primal scene associations which frequently feature beasts and fantasy creatures. It "terrified" Mary who would see in her bedroom at night "a man changing into a bear from his feet up":

> A big man-sized bear who almost seemed like to the ceiling standing in my bedroom and he'd *always* be in the same place in the corner of my room. It was the process of metamorphosis that was frightening. He changed from the feet up. But that part of him was bear and part of him was man. And then the man bit disappeared and the *man* bit arrived.

I have emphasised the interesting parapraxis where the man disappears to be replaced by a *man* rather than a bear indicating some slippage between the two. Traditionally the bear is "the beast that walks like a man" (Warner, 1995, p. 301).[4] Similarly, Freud's (1918) patient Wolfman feared the wolf that was "standing upright and striding along". Wolfman's fear, Freud concluded, concerned his father and arose in response to his witnessing of a primal scene. I suggest it is plausible to link Mary's father to the bear, particularly as narrative events in the text are set in motion by a returning father, and Mary's father was frequently absent, at work or meetings. The fact that the bear changes from the feet up is another possible link to the primal scene, as this would be the way in which the scene would be encountered initially before the child's presence is noticed. This was also the position of the "lady" who sat at the *bottom of a bed* with her eyes pecked out in *The Birds*. Freud also analysed the symbolism of feet in relation to sexuality, as the child will often experience looking at genitals from below moving upwards. Also it has been noted that there is a relationship between narratives involving animal grooms, such as the bear, and the acceptance of sexuality (Warner, 1995; Bettelheim, 1976).

It is however more complex than just associations with sexuality. Mary explains the narrative:

> The princess was spiteful and selfish and full of her own importance and the prince was very lovely and kind and just didn't get back in time with this tree and so he got turned into a bear. And she had to learn to love somebody other than herself.

The omnipotent and narcissistic princess has to be displaced and learn to love somebody other than herself. Mary had been her father's "princess" and her centrality with both her parents is lost at this time, due to the arrival of her siblings. This representation of the father-daughter relationship in popular culture often revolves around this idea of daughters as princesses or fantasy creatures such as fairies (Walkerdine, 1993). This narrative features a prince who fails to get back in time, as well as an absent father, rather like Mary's father who was often absent and late home. This prince, who was "lovely and kind", is condemned to become a beast because of the self-centred actions of the princess who, in consequence, loses her beauty. This, the bear/prince tells her, is because the animals see her "true" nature: selfish and spiteful. The princess has to learn humility and kindness in order for her outer beauty to return. Consequently, it can be asserted that the bear is also a reminder of her own omnipotent, destructive and aggressive phantasies possibly towards her much adored father, which in turn precipitated some guilt that is characteristic of the Kleinian depressive position. Volkan and Ast (2014) have also argued that anxiety and conflict caused by sibling rivalry and also the primal scene can take a symbolised form and it is common for this to be in the shape of animals and birds.

The complex imaginary work associated with *The Birds* and *The Singing Ringing Tree* can be seen in the context of negotiations made as part of normal psychosexual development that takes place in relation to unique family relationships and circumstances.

Intergenerational identification

Sue: brothers, sons and Christopher Eccleston

This section will explore themes around mothering which appeared in Sue's and Mary's interviews, focussing on intergenerational identification and conflict associated with identity in transition. This will illustrate the way in which familial relations are lived psychosocially and affect the interpretation and meaning of favourite texts.

In the interviews Sue frequently talked about her brother Paul and her eldest son Jack. The stories were characterised by recurrent descriptive phrases of both as "hard work", "intense" and "difficult to live with". These discursive themes appear again when Sue talks about her secret passion, Christopher Eccleston, who is "known for playing intense, dark roles" that are "full of angst". She believes "he's probably like that in real life", indicating some blurring of the boundaries between fantasy and reality. Despite Sue initially not wanting a brother, once he was born, even though he got all the attention, she "adored him and still does" becoming a "little mother to him". Paul has been a cause of much anxiety for her mother who worries about his safety since leaving home to go to university in London, where he has remained and now has a good job. Sue also worries continually about her eldest son's welfare and his being involved in a violent attack or being killed, suggesting some intergenerational transferences of anxiety. In the midst of this, I want to show how Christopher Eccleston is interwoven into Sue's fears, anxieties, desires and phantasies indicating a connection between the three figures.

Throughout life, identity is in process, being restructured in response to change which can often be conflictual and anxiety provoking (Erickson, 1959; Faimberg, 2005). Mothering is one such change – a momentous life event where, for the first time, the woman has access to multiple identifications, with her own mother as mother, to her own experience of being a mothered child, along with the numerous incorporated internal object relations with significant others such as fathers and siblings (Hollway,

2009, 2010). Issues of similarity, difference, separation and ambivalence are emphasised in sibling and parent-child relationships and play a role in identification, as the reader will see.

Our Friends in the North (1995, Dir. Simon Cellan-Jones, BBC)

This drama spans a 30-year period and follows the lives of a group of friends from a working class community in the North East of England. Initially, the text had an aleatory factor taking Sue by surprise as she found herself gripped by the first episode and also by the strength of feeling she had on this first encounter with Eccleston. Sue's description of Eccleston is interesting as she explains that for her: "he's got a Geordie accent. Yeah, I mean he's an *actor.* He can't possibly have a Geordie accent but *that's* what *I always* hear." It is true that Eccleston does not have the Geordie accent that Sue perceives, but this, I suggest, points to an element of fixity associated with aleatory qualities of this text which resulted in Eccleston becoming an evocative object, which as I have noted is linked to the emergence of unconscious unthought-known knowledge. Let us think more about what this might be.

This text clearly provides opportunities for narrative identification for Sue, but it is her unique sibling experiences, in particular the observation of a younger brother's trajectory in a time of greater opportunity for social mobility, that crystallises around the figure of Eccleston in *Our Friends in the North*. Eccleston plays *Nicky*, a student, whose girlfriend, *Mary*, is also destined for university. Following a brief relationship with another member of the group, *Mary* becomes pregnant and marries, giving up her chance of going to university. Education, as a major route out of working-class life, is explored in the text and, for the character *Mary*, this revolved around the idea of throwing away her educational opportunities for marriage (which does not work out) and children. It begins in 1964 and ends in 1995, following the experiences of characters that are approximately the same age and from working-class backgrounds like Sue. Covering the 30-year period Sue has lived through, the narrative speaks to Sue's experiences; "marrying young", "babies" and her tenuous attempts to hold on to education, through night school and an unfinished degree. There is also the character of *Geordie*, the fourth member of the friendship group who moves to London, initially enjoying success, but eventually ending up involved in drugs and crime, capturing the cross-generational anxieties revolving around Sue, her mother and brother who moved to London, and her own fears for her eldest son.

In the various case studies, I have shown how families and the world of internal object relations permeate viewing. Sue employs a combination of projective identification and unconscious phantasy where Paul and Jack become intermingled in the figure of Eccleston/*Nicky*. This is possibly because of a combination of physical similarity coupled with narrative identification in the aleatory moment. As an evocative object Eccleston contains the complicated "intense" and "difficult" parts of her brother and her son. This is where the internal meets the social, however, as Paul, like Eccleston's character, was able to go to university, live in a large city and become an attractive cosmopolitan man. Sue addresses her own life choices through the differences between her experiences and those of her brother, who would have been at university at the time the series was transmitted. This was a difficult time for Sue, and she describes feelings of resentment towards her brother:

> We went through a phase, I suppose when, when he was away at University and I was so young when I got married, and I'd got the kids. And I remember thinking, "maybe I should've done that, maybe I should've gone that way". And there was a bit of resentment that he'd done it and I hadn't. But it passed. We did have a bit of a clash about that time.

Paul's trajectory confronts Sue with the reality of her choices to marry and become a mother at a young age also seen in the narrative of *Our Friends in the North*.

One effect of siblings as a set is that each member must struggle to assert their individuality and difference within the set. For Sue, the 12 year age gap meant a longer than usual separation between their respective moves into education. Their differences were, in part, imposed upon them by the reality of changing socio-economic relations, in particular, as we have seen in Chapter 3, widening participation in education. Sue recognises she might have been a different individual if she had made different choices. She is also aware that the choices she made were gendered in that she did what most of her female friends did at the time. This illustrates the psychosocial nature of identification as her internal object relations and phantasy life, formed in her particular family circumstances, meet the social and cultural realities of gender and social class.

This can be seen again in another Eccleston/Paul text featuring education as a site of struggle: *Hearts and Minds* (1995, Dir. Stephen Whittaker, C4). This text was discovered around the same time as *Our Friends in the North*, where Eccleston starred as a teacher who is:

> full of angst. I think he'd been a miner and retrained as a comprehensive schoolteacher, and he was a *brilliant* teacher. *But* all the other teachers hated him because he was so upfront, so outspoken because he'd come into it late. I mean he had fights. I think he had a *physical* fight.

Clearly, there are issues around re-training and coming to the profession late in life that would have been evocative for Sue who has made a number of unsuccessful attempts to return to education. However, there is a narrative about Paul, offered later in the interview which adds another layer to the evocative nature of this text providing an example of the way unconscious links are made between texts without being consciously articulated. Sue returns to the subject of Paul and explains:

> He's so *intense*. He's got himself into a few scrapes at school, erm a few clashes with teachers. Because he's quite erm tall and handsome and he grew up really quick. When I say grew up he became a young man quite early and he grew up. I know he had one or two clashes with male members of staff. There was one teacher in particular that he did have a sort of "strutting antlers", they used to lock antlers. They went on this school trip and this teacher was there and they actually, literally, they *physically* came to blows.

In addition to the similarities in content such as the intensity and angst associated with the two men and the physical clashes in the school, this is also a narrative that suggests conflict associated with "growing up early", something Sue has experience of when she became a "little mother" at 12. Despite *Paul's* problems at school, unlike Sue, he made it to university and has a career; this would explain why they "did have a bit of a clash around that time".

The theme of education and social class as sites of conflict appears again in *Jude* (1996, Dir. Michael Winterbottom),[5] another Eccleston film Sue encountered when her brother was at university. Sue chooses, however, to focus on what she sees as a troubling feature of the narrative. She explains that she thought the two main characters that have a love affair were "related" and that this is "quite controversial", particularly as she remembers them as "siblings". I explained to Sue that the characters were actually cousins and she reassures herself saying: "yes they must've been cousins they can't have been *siblings* can they?" Which would, of course, prompt incest taboo anxiety if, as I have proposed, there is some conflation between Eccleston and Paul. Conversely, Klein (1960) examined the role of siblings in psychosexual development, pointing to desire for siblings taking place in phantasy as a feature of normal development. Sue is attracted to Eccleston who is:

> often usually involved in sex scenes. He gets his kit off. I've seen his bum loads of times. There's a scene I can remember where he's on the stairs with his girlfriend and that's fairly typical. I'm getting all hot [actions, pulls at top and waves hand as if cooling face] [Laughs] and I'm *really, really* jealous of his female co-stars.

The desirous sexual elements cannot be divorced from the partially-fused identification between Eccleston and her brother. As Sue approached puberty she was faced with the reality of parental sexuality, which became located around the fact of her brother, so it is not surprising that Sue should associate awareness of her parents' and subsequently her own sexual awareness, with her brother. More than this and following Coles's (2003) argument, it also points to the contribution of sibling desire in mature sexual fulfilment. Sharpe and Rosenblatt (1994) have also pointed to the existence of Oedipal sibling triangles being as significant as and parallel to parent-child Oedipal relationships. Similarly as Oedipal conflict can continue to emerge throughout life, Sharpe and Rosenblatt note that sibling Oedipal desirous configurations also remain significant and are reactivated in response to life events (see also Volkan and Ast, 1997).

To understand and draw together some of these themes we must return to the film *Big* and the moment, and it is just that, a moment, which always moves her to tears; when the adult *Josh* turns back into the 12-year-old boy.

Josh's adult girlfriend, with whom a sexual relationship has been implied, has driven him home and he is left wearing his executive suit, now far too big for him. He leaves the car and walks towards his home and mother. This is how Sue tells the story:

> It's just the way he's lost his innocence, do you know, he's lost his innocence but he's come back, and he's gonna grow up and go through all that but still he's lost his innocence. I mean, well, he's

> basically, he's been with a *woman*. But I think that, he turns round and the *mum* (lots of emotion in voice) oh it's just like your baby's come back. And then when *she walks* [my emphasis] through the door. When he walks through the door. You don't see it, well you don't hear it. You just see his girlfriend looking through the rear windscreen. And he goes out as Tom Hanks in his suit, walking towards the door then she looks again and he's got this *huge* suit on and he's gone back to his original size. And you don't hear anything. You don't hear the mother say "what on *earth* are you doing in that suit" *she* [my emphasis] just goes in and she's so relieved.

Here we can see how Sue's inner world and biographical experiences make this a particularly evocative moment. Sue makes the link that I had previously proposed that the loss of innocence associated with sexual relationships was traumatic. At the time of release onto DVD/VHS, Sue's brother would have been around 12 or 13 years old, the time at which Sue was forced to confront sexuality. There are also identifications as a mother which would come into play as Sue's eldest son would have been a small child. Identification shifts in this short scene, demonstrated in parapraxes as Sue twice says "when *she* goes in" indicating her own identification with this significant moment of transition. At the same time she identifies with *Josh*'s mother who gets her baby back, something Sue did in response to her daughter going to secondary school when she had her third child. There are also the feelings she shares around her brother and son. Overnight, *Josh*, like Sue, grows up and becomes a 30-year-old man, and this would be the approximate age of Sue's brother at the time of the interviews, who also "grew up" too fast as we saw in the story about his schooling. She would also have seen *Big* for the first time after becoming a mother to her own son. In this moment, there are a range of intergenerational and cross-gender, sibling, parent and child identifications which come together to make this a powerful emotional moment.

Mary: both daddy's girl and daddy

I have followed Coles's (2003) argument that sibling identification is often more complex than the traditional Freudian models which focus primarily on hostility (see Mitchell, 2003). However, Mary's identifications and responses following the arrival of her sisters appear to follow this classic model. Apart from discussing the impact of their births in entirely negative terms, Mary's siblings are noticeable by their absence from her discussion.[6] It is her relationship with her parents – and her father in particular – that structures the interviews and elucidates the role of families in producing gender identifications diachronically.

Mary remembers watching *Stingray*[7] (1964–1965, Dir. Gerry Anderson, ATV) as a child and being fascinated by the mermaid Marina who represented traditional femininity:

> She was very floaty and girly and feminine and princessy. With long blonde hair and floaty dresses ... there was a lot of fluttering of her eyes and she was so lovely underwater. And girly. And she didn't want to wear the trousers, going on missions, she was happy being a girl and that was nice.

Mary also makes clear this was a form of dis-identification (Skeggs, 1997) as she was part of a trio of girls who were expected to conform to cultural expectations of femininity and "made to be girls":

> I wasn't allowed to wear trousers. When I was little I *always* had to have a dress on. I wasn't allowed to have trainers. I had to have *proper* shoes. We weren't allowed to have train sets or that because *they* were *boys' toys*. So probably I've always looked for strong female role models rather than, well because I always wanted a train set[8] when I was little. So maybe I equated that sort of thing with women who weren't all pink and fluffy and women that were more strong.

The fact that Mary felt femininity had been enforced upon her, coupled with her displacement as the only child by two female siblings might prompt some antagonism towards being a girl. As part of the process of sibling differentiation, Mary identified with masculine characteristics and her father. This was reinforced as Mary stepped into her, often absent, father's role for her mother, evidenced in her mother's adult conversations with Mary. This can be explored further through Mary's relationship with her father and her son and the film *Father of the Bride*[9] (1991, Dir. Charles Shyer).

The film is a sentimental comedy centred on a father's struggle to come to terms with his little girl becoming a woman, leaving home, and his replacement in her affections by another man, her future husband. As the bride prepares to leave the wedding reception, her father desperately tries to have a moment with her alone so he can say goodbye, but he is thwarted at every attempt and ultimately fails. This moment unexpectedly reduced Mary to tears because, she explains, the scene "mirrors" her own wedding. This was an aleatory moment which captured past and present experiences and conflicts:

> When I got married *I physically couldn't* say goodbye to my parents. At the wedding reception I said goodbye to everybody else but I physically couldn't say goodbye to my parents. I managed to say goodbye to my Mum but I *couldn't* say goodbye to my Dad. Me and my Dad have a very close bond. The scene at the end where she (bride) comes down the stairs. Then I became quite *desperate* to go before I had to face my Dad. And we got in the car and as we got in the car I said obviously a bit louder than I meant to say, "quick we need to go before my Dad comes". And I just remember an ethereal voice saying "your Dad's right here". And my Dad was just behind the car door. And I just *burst* into *floods of tears*. And we had to go away. I've never said goodbye to my Dad from when I left to get married and I never could.

Who or what was Mary identifying with? She describes the film scene as "gut wrenching" and "I couldn't believe how close it was to how I felt". If we examine the narrative we find that the bride was happy and did not have any problems parting from her father; in fact the grief and anguish in the film were actually located with George, the father. The film, she says, "mirrored a moment when I felt very vulnerable", but the vulnerability in the text rested with George. The film certainly triggered a memory where she felt vulnerable as a daughter, particularly around the issue of separating from her father, but it is as a *parent* that she now feels vulnerable. Whilst object relations often concentrate on the difficult passage of the child's separation from its mother, less is said about the difficulties the parent faces of separating from the child. Weddings, like births, are key moments of change in identity, but they are also poignant flash points for the reactivation of separation anxiety. This aleatory film moment is so intense because it encapsulates intergenerational identifications between parent and child that Mary faces at the time of the interviews: "I'm kind of facing that problem the other way round now as my son is due to leave and get married and I'm not quite sure how to let him go." Mary speaks as both bride *and* father of the bride.

Sharing this narrative, however, prompted a re-evaluation of the events and relationships:

> My Mum and Dad did a lot of work for my wedding but at the same time they were very controlling as well. So they weren't very much

> like Steve Martin in the film. In a way there's nothing in the film that resembles our wedding.

It is interesting to note how the scene went from something that emotionally "mirrored" her wedding, to one which bore no resemblance at all and pointed to the relationship with her father as one of ambivalence: "everything had been a big battle particularly with my Dad". Despite this she recognises her identification with him: "My Dad likes to be in charge. I've probably inherited a bit too much of that from him."

Intergenerational identification is also a feature of another narrative which gives an indication of the intensity of feeling separation and change precipitates:

> I was telling Sam not long ago. Now *I'm* facing the prospect of him leaving. I can remember the night before I got married. I remember hearing my Dad – you'll understand what I mean but the only word I can use is *howl*. It was a cry like an *animal*. And I can remember him just *sobbing* in his bedroom and me not daring to go in because I couldn't face it. I understand how it was. Because often I find myself in tears when I think about Sam leaving. I know it's gonna break my heart when he goes.

It is interesting to note Mary's use of animal motifs in her description of her father as a howling beast, recalling the creatures such as snakes and bears that stalked her in her bedroom as a child. The intensity of her relationship with her father is something that characterises her relationship with her son which also manifests intergenerational and cross-gender identifications:

> We're far closer than a mother and *daughter* [my emphasis] should be (*corrects herself*); a mother and son should be. We're like best mates, sometimes I think that's *bad* but I wouldn't want it any different. But you kinda realise you pay the price for it at times. We pay the price of – not to put a strange twist on it – but of *adoring* him so much. And I mean we just *do*. That was the reason we never had any other children because we could never imagine sharing him with anybody. But I just *knew* from the first second I set eyes on him I couldn't *ever* share him with anybody. And it was just a very conscious decision from then on that we wouldn't have any other children.

The parapraxis I have highlighted encapsulates two points of identification for Mary between parent and child: the daughter of her parents *and* the parent of her son. The story of her father howling with grief points to the intensity of the relationship which previously Mary had experienced as daughter. Now she experiences a similar intensity in her relationship as parent to her son. The decision to only have one child preserved the exclusivity and intimacy between the parent and child in an oedipal configuration (Faimberg, 2005, p. 50), which also remedied the displacement caused by siblings. Mary has an only son who can "wear the trousers" and play with train sets and not be displaced by any other females, a position Mary phantasised about as a child. She married a man who was the only boy in a large family of girls, which he hated. Thus, here is someone who shared her experience of competing with sisters and at the same time was comfortable with allowing Mary to "wear the trousers" and be in charge – as she notes of her inherited characteristics from her father.

In the extract below we see another shift in identification as Mary reverts to the little girl who disappoints her father. She is able to address this, though, through a temporary shift into her son's position, to please her father in a way she had not been able to do as a young woman. This is evidenced once again through a parapraxis:

> I was really pleased when Sam got his degree results and I said to him "ring your Granddad". I was really pleased that he was able to ring my Dad and tell him. I knew that my *Granddad* [my emphasis] would be just over the moon about it. I was really glad that *I* didn't ring my Dad because I'd be doing it for all the wrong reasons, to impress my Dad. And I know that he absolutely adores me and he thinks the world of me but I've always felt kind of felt like I'm not quite good enough. And I've always felt like just doing one more thing that will make it all right when I know that that's not true and I feel very much a little girl when I have an argument with my Dad. I couldn't tell him I didn't agree with him without feeling really quite bad about it. I mean I love my Mum to bits as well but I'm definitely a Daddy's Girl.

The discussion moves once more to *Father of the Bride* which she again uses comparatively to explore the difficulties of her relationship with her father. This extract also points to the intergenerational transmission of internal conflicts:

> I would find more reality in *Meet the Fockers*[10] because my Mum and Dad were very much "Oh dear she's going to marry a builder".

> When Phil asked if it would be all right if we got engaged, my Dad's comment was very off hand and he said "Oh I dunno what you're asking me for" and I was very disappointed in that, 'cause I know how much courage it had taken for Phil to do that. I was really angry with my Dad for not giving it the respect that it deserved. I'm quite sure that they felt it was all a bit beneath me and that really annoys me because my Dad very much had that from my Grandparents. My Dad very much felt that for her parents, he wasn't good enough and he found that quite a battle and I would've hope he wouldn't make the same mistake.

The "end" result of the child's move through the Oedipus complex, the identification with the "correct" male or female parent, and the separation issues involved in becoming a differentiated subject, are major achievements, but as this study suggests, they are also unstable. This is why it is valuable to extend the idea of the Oedipus complex into an Oedipal configuration (Faimberg, 2005, p. 50), which acknowledges intergenerational patterns of identification.

Walkerdine (1997) argues it is not enough to explore the child's desire in relation to the parents; the phantasies of the parent about the child are equally important. The child's identifications are formed through projective mechanisms and in relation to specific parental narcissistic conflicts that may have been inherited as the extract suggests. Narcissistic identifications are also the result of the way in which the child is "approved" or "disapproved" in the family, in historically specific circumstances (Layton, 2004). The parents, often as a result of some inherited narcissistic conflict of their own, project omnipotent characteristics and phantasies onto the child, which the parent wishes to see reflected back to ensure the continuation of the omnipotent phantasy. (Faimberg, 2005, p. 18). The child becomes "instructed by the implicate logic of their unconscious relational intelligence in the family's way of being" (Bollas, 1992, p. 51). At the same time, transference within families contributes to the development of familial "shared constructs" which provide unconscious templates to family members as they approach reality and the social world (Reiss, 1981). For Mary (and her sisters) this included the pressure to perform a narrow form of traditional femininity, attain upward mobility through educational attainment and be a "good girl".

In most cases, it is impossible for the child to meet the narcissistic desires of the parents, who can only love the bits of the child that acknowledge those loved or approved parts of the parent's self. The risk is that the child, fearing disapproval, responds by only expressing those aspects of the self that display the approved aspects of the parents: a form of false self. The child must not only collude in the parents' narcissistic phantasies but he must also become a repository for hated and unbearable parts of the parents' own history (Faimberg, op. cit.). We can see above that Mary and her husband are carrying historical traumas for her father who also suffered from disapproval and feelings of social failure. Mary carries the unbearable "not good enough" part for her father; something which emerges when the child tries to separate and articulate his spontaneous true self. This process explains Mary's attachment to layers in texts where characters hide parts of the self so "things are not what they seem". As a child, Mary had to relegate her own childhood and spontaneous needs to those of her parents. Developing a form of psychic self-reliance, however, she dealt with conflicts primarily from her own resources. This explains *Mary's* attraction to the "bad" girl characters Bea (*Tenko*) and Vivian (*Pretty Woman*), which enable her to make contact with her own bad girl objects, which could not be articulated, because they were completely unacceptable to her parents. If Mary had developed an outer false self it was successful at protecting the psychic vitality of the inner self and allowing her to eventually wear the much longed-for trousers and overcome the issues of displacement caused by her siblings' births.

The child must achieve separation from the parents but, at the same time, the parents must also differentiate their own histories and desires from the child, and this is a key tension. It is this Oedipal configuration that accounts for the points of slippage in Mary's identifications where she moves between daughter, mother, son, grandson, daddy's girl and daddy in what Faimberg calls a "telescoping of generations". In this telescoping the difference in generation is blurred, allowing Mary to experience the scene in *Father of the Bride* from a variety of positions.

This chapter has shown that patterns of identification are psychosocial, shaped by life experiences, relationships and the projective and introjective traffic between inner and outer worlds. The aleatory and mnemonic textual moments I have explored show how viewing can be used imaginatively to address the conflicts and difficulties associated with identity in transition including separation and change. Complex forms of identification with textual moments, characters and narrative have shown that siblings play an important role in the inner world and in psychosocial processes of separation and individuation. The case studies have shown two different models of sibling relationship. The traditional or "hostile" model views the arrival of siblings, for older children in particular, as one of intrusion and rivalry, focussed around displacement and the loss of a privileged relationship with parents, which captures

Mary's experience (see also Chris, Chapter 3). This contrasts with the model that views sibling relationships as loving, desirous and concerned with separation and individuation issues seen more clearly in Sue's case study. Complex forms of identification with textual moments, characters and narrative have shown how the arrival of siblings, with the accompanying traumas associated with sexuality and the social and psychic consequences that follow these, were experienced. Siblings do not only have an effect on the world of internal object relations, they are also important "external" objects against which life choices and trajectories can be measured. Evidence was presented for the way the life choices made by two typical working-class women born in the 1960s negotiated their life choices to marry young and have children, and how viewing and favourite texts play a role in this psychosocial identity work, which is something that is missing from existing film and television audience research.

6

Risky Viewing and Risky Method?: Psychoanalysis, Method and Defended Viewing

Previous chapters have employed psychoanalytic concepts and ideas to understand identification and subjective engagement with viewing. As many researchers working psychosocially have embraced the use of psychoanalysis to inform research design and analysis, it is important to ask how these developments and observations translate in practice. Given that this book is a study in the psychosocial, this chapter will concentrate on the use of psychoanalysis and method. It will focus specifically on the idea introduced in Chapter 2 of researcher subjectivity and unconscious communication as resource and will explore this through an account of the unconscious dynamics of a research relationship.

The relative novelty of adopting psychoanalysis for empirical projects means that researchers face significant challenges in designing their interventions. These issues revolve, as I have noted, around an important ethical question: the transposition of perspectives and techniques designed for a specific location (the clinic) and purpose (to aid in the treatment of emotional difficulties) into a research situation for which they were not originally designed. However, this shift could not take place at all unless there were some clear similarities between qualitative research and psychoanalysis. They are both interpretive endeavours – the difference lies in their aims. Specifically, interpretations are made in clinical practice in order to provide the interviewee with an understanding of his or her unconscious emotional difficulties. In contrast, it is unlikely that interpretation would take place in a qualitative interview situation where, once the process of interview is complete, the data is subjected to systematic coding and thematic analysis.

To begin, it is worth noting that the interview space and the consulting room may both be viewed as contrived encounters which bear no resemblance to a natural communicative exchange. Whilst psychoanalytically informed research does not attempt to transpose the psychoanalytic clinical template, it is useful to note that *similarities* exist between the two settings. They are both concerned with talking and listening; where one person is trying to make sense of the experiences that the other participant is trying to communicate. This chapter will also examine the contribution psychoanalytic theory can make to understanding communication between interviewer and interviewee, hence the idea of presenting an account of the communicative dynamics of a research relationship. Understanding communication more fully, including the recognition of unconscious processes, is not necessarily ethical in and of itself, but it does create the foundations for a more realistic appraisal of data production and its analysis. Similarly, as we saw in the study discussed in Chapter 2, if researchers genuinely want to *hear* what their interviewees are saying to them (be it consciously or unconsciously) they must be alert to their own desires, particularly concerning the demand to produce data, their need for omnipotent control, and the ways that their own anxieties might impinge on the research process.

This chapter features the case study of Bill,[1] a recently retired teacher, who is middle class and married with a grown-up family. As the reader will see, the value of this case study is that it encapsulates some of the tensions that can occur when psychoanalysis is used in non-clinical qualitative research situations. It offers, therefore, a specific opportunity to examine the practical, ethical and conceptual use of psychoanalysis as method, as well as continuing the project of evaluating its interpretive utility for audience studies. The case study detailed below presented some difficulties, as researcher and interviewee had different aims (see also Chapter 2). These tensions can, however, be used to examine the role of unconscious conflict in communicative difficulties from a psychoanalytic perspective. This chapter returns to the presence of anxiety, phantasy and the idea of the defended subject as *both* researcher *and* research subject. It also explores the idea of defensive spectatorship.

Conflict, as we have seen in previous chapters, is not only generated intra-psychically in the individual but is also produced by intersubjective processes: feelings, affect and anxiety are the result of exchanges between human subjects, which means that greater attention must be paid to the role of the researcher on what is, and what can be, thought and said during the interview. Interpretations emerge, therefore, from the research relationship and *not* the individual mind of one person (Ogden,

1999; Symington, 1988), and in keeping with the spirit of free associations feelings, affects, dreams and other mental images act as forms of evidence. As we saw in the Chapter 2 interview, situations can generate unconscious conflictual processes and anxiety in both researcher and interviewee and, for this reason, interviews should be approached with an awareness of the presence of anxiety as something normal and inevitable. The difficult research relationship provides an opportunity to narrate processes of thought and interpretation, which is essential, if the researcher is to make claims about the accuracy and validity of interpretations and conceptual generalisations. In turn, this also represents a novel contribution to the field of communication studies.

The chapter is organised in two sections; the first takes up the idea of researcher reflexivity and defended subjects (Hollway and Jefferson, 2010) as a prelude to the presentation of the case study itself. This section was inspired by the clinical vignette, where the clinician reflects on their feelings and any difficulties that arise during the process of interpretation (including the role of his or her own subjectivity in arriving at an interpretation). Whilst the second section presents a heuristic account of the way I worked through a number of challenging methodological issues, reflecting on practical and ethical concerns. The chapter contains, therefore, an expression of the author's own understanding of the emotional and interpersonal aspects of the research relationship. In adopting this focussed but idiosyncratic approach, I share Symington's (1988, p. 11) hope that the reader can cross-check their understanding against my own, on the grounds that one's "own self-understanding is enormously enhanced by coming up against the personal understanding of another". This contributes to current debates about researcher reflexivity and subjectivity in research practice.

Researcher reflexivity, unconscious communication and the "defended" researcher

Much debate has taken place in recent decades about the role of the researcher and the researcher's "self" in the process of interpretation (Olesen, 2000). Some theorists, (Frosh, 2010; Frosh and Baraitser, 2008) whilst sympathetic to psychoanalysis, have been critical of attempts by researchers working with the FANI method, to adopt clinical concepts such as transference and counter-transference to understand communicative and affective processes in qualitative interviews. It has been suggested that these concepts are *unique* products of the clinic and cannot be transferred to the research setting. Where this shift occurs, there is a danger of the concepts being used inappropriately in the interpretation of the interviewee's unconscious anxiety. This perspective holds that these concepts are based on the development of a special clinical and therapeutic relationship between analyst and analysand, where the latter transfers feelings about figures from their past onto the analyst, and ultimately it is this distinctive relationship which brings about relief from neurotic symptoms. There is also debate about the practicality of these concepts in understanding anything other than neurotic difficulties which are of no interest to most qualitative audience researchers. Taking up the idea of the defended subject, another criticism concerns the attribution of anxiety to the interviewee. It has also been argued that non-clinically trained researchers fail to recognise that the anxiety they attribute to the interviewee is actually an unrecognised emergent feature of the *interviewer's* unconscious anxieties and desire (Frosh and Baraitser, 2008), particularly for the production of relevant data, which I pointed to in Chapter 2. It is this point I want to take up as I explore anxiety and the idea of the defended *researcher*. I also want to explore further what researchers might do with the feelings and thoughts that research generates.

The trend of researchers being reflexive about their social characteristics has been taken further by those researchers who use psychoanalysis. Hunt (1989) was one of the first to propose that the researcher's mental experience mediates their understanding of the cultural and psychological world of subjects. Freud recognised the fact that his feelings, thoughts, fantasies and dreams were complex forms of communication with his patients. He was certainly aware that self could play tricks, but it was his belief that, far from prejudicing the interpretive process, (prejudice being taken pejoratively) "these skills and qualities can be employed in the service of facilitating research" (Freeman, 1989, p. 307). We might note that recent neuro-psychoanalytic research has found evidence linking the right hemisphere of the brain to *implicit*, unconscious and affective information processing and the left brain more *conscious* processing. (Happaney et al., 2004): thus providing empirical evidence to indicate the human brain as responsive to, and generative of, unconscious communication. This offers a rejoinder to the idea of transference as unique to the clinic, as well as demonstrating the presence of unconscious factors in the conduct of all communicative encounters (Schore, 2010, 2003). Nonetheless it was Freud (1916, p. 194) who was the first to argue that "it is a remarkable thing that the unconscious of one human being can react upon that of another without passing through consciousness" (Freud, 1915, p. 194). However, as I have noted, there is something rather peculiar and contrived about research interviews that might predispose the relationship to a particular intensity, which I will explore further in this chapter.

Recent research also reflects a growing interest in other forms of unconscious communication where physical presence is removed, such as on-line communication and, as I discuss below, through letters (Suler, 2009). There is emotional messaging encoded in the communication itself despite the fact that there is no physical dimension to the interaction (and which may contradict the emotion suggested overtly by the written language). We still feel, fantasise, dream about and respond emotionally to other forms of data such as letters or email correspondence. Our responses to our interviewees and what they tell us is also on-going, we live with our feelings and we carry the interviewees around with us. For this reason the research diary is an essential tool in which researchers can record emotional responses over a long period of time. This helps to address our emotional involvement as researchers and assists us in thinking about the relationship between the anxious defended researcher and our methods of interpretation. This is a positive move in ethical terms, where it has been argued that researchers should not be seen as neutral vehicles for representing knowledge. From a practical perspective as well, research should be both fair and democratic, and researchers should also be willing to offer up their own accounts of their experience to interpretation and scrutiny (Hollway and Jefferson, 2010, pp. 3–6). Following this call for researchers to scrutinise their accounts of the research experience, I base much of the discussion in this chapter on my use of extensive research diary entries.

This case study is of particular value because it allows for an exploration of the idea of the defended *researcher*. Prior to interviewing Bill, I had conducted interviews with the other respondents, which had been productive and generated what I perceived as rich data in relation to the research questions about emotional significance of texts. As I will explain, in Bill's case, he had his own agenda for the interviews which then forced me to confront the fact that I might not generate data that would contribute towards answering my research questions. Along with this, he also raised ethical issues about the method, making my work with Bill doubly defensive on my part which I will explore below.

Bill was another interviewee who had experienced social mobility. He grew up in a working-class home and won a scholarship to a grammar school and later went on to study at a number of elite institutions: like Chris who we met earlier, he is another cineaste, whose approach to cinema (he "rarely watches TV") is one of intellectual appreciation.

Most of the films he chose to discuss came from, or were first encountered at, a particular moment in his life; his late thirties and early forties. He remembered this as an exciting time but also that his identity had undergone some significant changes including a major career shift as he moved into teaching. He had also married a woman his father disapproved of, after which his relationship with his father began to "unravel".

On volunteering to be interviewed, Bill at the outset generated a question about anxiety and viewing, stating that he found cinema spectatorship "risky". It became clear that Bill had volunteered for the project because he had his own agenda. He used the maxim "know thyself" to explain that he wanted to use the interviews to "learn about himself". As well as cinema being a "risky" activity, he also described the idea of taking part in an interview, which he was going to use to explore his identity, as both "exciting" and "dangerous" and that this added "to the attraction of taking part". This forced me, as the interviewer, to confront something that I had not previously considered: that his aims might potentially conflict with my own and that his desire to explore "risk" made *me* feel anxious. It was also not inconceivable that I might ascribe this anxiety to the interview subject as noted above.

When I had conducted previous interviews, I had felt the need to exercise care in my treatment of research subjects, falling back on tried and trusted policies surrounding confidentiality, withdrawing from the interview if need be, destroying data when appropriate or stopping the interview if requested. Bill's wish to undertake a "risky" exploration of his identity posed an unanticipated ethical dilemma. I had no idea where Bill would go with his journey and what my role would be. I was forced, in the throes of an interview session, to consider the importance of my own investment in being *in control* of the interview. This clearly pointed to issues of power within the relationship. In Chapter 2, one of the tensions described in the discussion of the troublesome interview concerned the researcher's educational superiority over their interviewees. Here my anxiety concerned a highly intellectual interviewee, whose intellect was certainly on a par with my own, who might possibly be able to take control of the research in ways that made me feel anxious.

Bill not only had an agenda of his own but he also brought *the explicit language of clinical psychotherapy* to the table. He said, for instance:

> Your method slips into some sort of therapy, whatever your intentions *or* mine. I can feel an internal censor restricting what I can say.

> First of all, from my point of view as an interviewee, I cannot assume you are competent as a therapist. Then, from your point of view, I must beware of dumping stuff on you, with all the transference and counter-transference that that could involve. I cannot assume you have the support system I would expect a therapist to have.

It would be rare for an interviewer, in most types of research, to talk to the interviewee in a technical language, even if this language might be used later when data was written up. I was unsure if Bill wanted me to interpret what he said in a clinical language which was, of course, not possible, as the method was psychoanalytically informed and not strictly clinical. Bill also presented me with a rebuttal of the anti-defensive design features of the FANI method (he said he was *defensive* and hence restricted in what he would say). He identified the more problematic, clinical view of transference discussed previously and also raised an important point about supervision. From my point of view as researcher Bill had turned everything on its head.

My initial fears about the interview were not realised once it was set in motion. Bill was a very pleasant and articulate man who talked enthusiastically about his favourite films and at the end of the interview said he "enjoyed himself". In relation to the task of self-exploration which Bill set for himself, it was evident that he was in an explicitly self-reflexive mode in the interview. He differed from other interviewees in that he employed the language and conventions of what could be deemed a prototypical self-reflexive *discourse*. He discussed his life experiences, exploring his biographical identification with narratives and plot lines in his favourite films: *Matewan* (1987, Dir. John Sales), *The Battle of Algiers* (1967, Dir. Gillo Pontecorvo) and *Distant Voices, Still Lives* (1988, Dir. Terence Davis). Bill used the films to discuss his identity explaining, for example, that *Distant Voices, Still Lives* reminded him of his changed class position, saying: "I feel in some sense I remain deeply working class, although I know that I could no more fit into that world than into the clothes I wore in my childhood." Bill was adept at talking about films and relating them to his identity. He used the films to reiterate, in a rather repetitive manner, what he called a "series of allegiances": to politics, liberation, teaching, multiculturalism, the working class and egalitarian relationships with women, thus presenting himself as a culturally sensitive, intelligent, left wing and pro-feminist man.

This, on reflection, seemed to fulfil the open-ended invitation to talk about the film and its emotional significance, but actually, in this case, it produced logical, analytical and intellectual answers. Each account of his identification with the film and its connections to his identity would be accompanied by an intelligent critical evaluation of the film, such as its "painterly aesthetic qualities", informed by ideas from auteur, genre and narrative film theory. Such a discourse would not look out of place on the pages of an academic film journal or textbook. Wood et al. (2008) note that self-reflexivity often appears discursively in interviews with middle-class subjects who, they say, are skilled at utilising a wide range of available cultural and linguistic resources. Self-reflexivity is a marker of middle-class autonomy and forms an important part of the construction of a middle-class identity through discourse. This certainly seemed to be the case with Bill. Psychoanalytically, couching experiences in a discursive form is also an example of the use of intellectualisation, which is employed defensively by the subject to master conflicts and emotions (Laplanche and Pontalis, 1973, p. 224). Clinically the analysand who uses intellectualisation presents as one who is open, offering dreams and emotional experiences to the analyst. Significantly, they are keen to offer their own interpretation, as Bill did, as part of a defence against the risk that interpretation might bring (Abraham, 1919, pp. 83–84). I discuss some examples of tensions between his interpretations and his wish for me to analyse him below. As Bill's identity was in transition, having recently retired from a job he enjoyed, facing the prospect of negotiating a new later-life identity and ultimately his mortality, it is not surprising that reflecting on the past might generate anxiety. The fact that he chose to focus heavily on a particular time period where he obviously felt secure in his identity would serve some defensive function. Where intellectualisation is in operation, it has been argued that, a form of "empty" speech occurs (Bollas, 1995) where a series of elaborate and eloquent words are used to keep feelings and anxiety separate and at bay. I want to explore this point further.

Immediately after the interview, my research diary records strong visual responses of beautiful and epic filmic images, with Bill as a central figure in the frame. Despite the powerful lucid images, I found I lacked *affective* responses to the data, immediately after the interview and over time, as I tried to engage with it. In previous interviews with other respondents, I perceived I was able to understand what was being told through feelings. I empathised and shared the experiences through embodied affective responses. For instance, I could *feel* Chris's twitchiness, Sue's fear for her son's safety and Sophie's love for her dolls and bears. In keeping with Matthews (1998) who notes that the body is a physical instrument which can also listen and attune unconsciously and emotionally with another being, with Bill I could understand cognitively, very clearly what he was trying to communicate to me about his identity and experiences, but I just *wasn't feeling* his perspective. The filmic visual responses I had were striking but they lacked affect, rather like the elaborate coat which covers the empty speech of intellectualised discourse. The manifest content was stunning, but the latent content was hollow.

As Bill had flagged up the theme of risk, I found that I kept coming back to, and trying to identify, exactly what it was that could be considered risky. Two extracts stood out; one was told in the interview, the second in the letters. The first concerned a first-viewing experience of what became a favourite film: *Matewan*. What had begun along the usual lines, based on an intellectual evaluation of a particular film's form and the idea of auteur, shifted in tone:

> I went with a female student from school[2] who had just put her kids and a few things in her car and run away from her marriage so she was pretty stressed. And I found her *very attractive* (interviewee emphasis) and sharing it with her... there was this additional buzz of that sharing it with her. [...] There was something anarchic and wildly romantic about her. She was extremely good-looking with very beautiful eyes. I was swept away emotionally. Eventually it became a *cooler, professional relationship* (author's emphasis) but I still have very warm feelings.

Here was a situation where Bill expressed feelings of desire for a woman who was not his wife, which was clearly an identifiably risky situation for him. On viewing the data as a whole, I later realised this stood in sharp contrast to the idea of what he referred to as his "safe", "celibate" relationships with women. This became relevant as the research progressed through letters.

Despite not wanting to impose my own agenda or ask questions that might be construed as therapeutic, I asked a question which, in retrospect, shifted the discourse in a direction I had not intended. In trying to work on the issue of reflexivity and the effect of the researcher, I asked if the similarities in age and situation between me and the student might have triggered this memory. Bill's response was a long narrative which began with another reassurance of his enjoyment of safe, celibate relationships with women, possibly for my reassurance, which then moved into a story about an older poet who he had met (who was slightly older than Bill was at the time of the interview, suggesting some identification with this man) and who had expressed how hard it must be to work in an environment (a school) full of young beautiful women:

> To be honest I see you as an attractive woman whose company I have enjoyed and I think of you as safe, being both married and young enough to be my daughter. I suppose it recalls safe but warm friendships with women. Students and colleagues in my career as a teacher. I remember once taking the poet (name) through the corridors of (school name) to the room where he was to give a reading. I realised I had lost him and went back to look for him. I found him standing transfixed amidst the crowds of young people on their way back to their classes. "Jesus Christ," he said. "How can you possibly work in a place with so many beautiful women?" He was then in his seventies I guess. I thought, I know these young women more through their writing and discussion than through their appearance. There is not much logic to this.

On the surface then, this is perhaps another narrative about young women and the risks associated with desire, which the poet sensed and articulated.

The second part of our research relationship was conducted through letters. As I have noted, more research is needed on the unconscious research dynamics of letters and digital correspondence, as once physical presence was removed a new dynamic emerged. The focus shifted from relationships with women, and Bill sent me a number of long written narratives concerning his relationships with men who had been colleagues and friends. Each letter abounded with narratives like the above extract which induced the logic of free association and solicited my interpretation, for example: "something came to mind from way back", "there's not much logic to this story", "I don't understand" and "perhaps it is your job to make sense of this". One such narrative was the second story that initially seemed to point to risk but, this time, to something concerning desire and men.

Whilst there may be uncomfortable issues around women, I sensed that there were also risks around men, which may have been sexual in origin. The following narrative about a childhood experience with a group of boys, "came to mind", although he was "unsure why" when he was thinking about his identification with working-class life in *Distant Voices, Still Lives*:

> I guess I was about nine I went up to x into the countryside from my hometown of x. I had felt safe in the company of the big boy from next door who was, perhaps, eleven. A gang of much bigger boys pestered and held us into gorse bushes. I then felt isolated but tried

> for a sort of self-protection by feigning to make light of it and trying to make the gang laugh. I can remember saying, "These prickles are pricking me." At the time I thought it was very witty and I thought it had some sort of sexual reference, which I did not really understand. Gosh, this story has come from a long time ago. I am amazed that I remember it. I shall desist in trying to analyse why that might be. Perhaps this is your job?

Clearly, Bill was keen for me to analyse him and he draws attention to the "risky" sexual undertones of the narrative around awareness of sexuality and possible sexual feelings around boys. This story is similar to those catalogued by Savin-Williams (2001, p. 117) where unusual moments occur in childhood which are not overtly sexual and then unexpectedly provoke sexual thoughts and feelings and can also be seen as an example of après-coup (Laplanche, 1992).

Bill's continued requests for me to analyse him because he recognised the psychoanalytic features of the method were generating feelings of anxiety in me. The requests continued in a letter that arrived after I thought the research had ended. Bill sent me a letter explaining that the "work" he had done with me (I presumed he was using this word in the clinical sense) had been "very exciting" and he wanted to keep working with me. He also explained that he had decided to begin psychotherapy. I reiterated that I was unable to do this and suggested some books on film therapy that he might find interesting and could discuss with his therapist (See Hesley and Hesley, 2000).

Following this development, I noted that I was drawn to Bill's words in the previous extract about the female student advocating a "cooler professional relationship" as well as some personal anxiety about the data. After various attempts to code the data, I found myself unable to connect with the material intellectually. I just could not make sense of it and found myself thinking that the only analysis that was worthwhile in relation to Bill's data was discourse or some other form of purely linguistic analysis. Eventually, I discussed this with my supervisory team and decided I was not able to use the material for a research project on viewing, as the data seemed only to reproduce the most obvious, well-worn ideas about identification with on-screen characters and plot, genre, auteur and established theories about film form and meaning. I made a decision to drop this case study and find a new interviewee to replace Bill.

Once I had let the material go and decided to move on, however, I found that during the next couple of weeks a change occurred. Bill's data kept coming into my mind. My research diary recorded a dream fragment. From what had been a still image, similar to the static visual beauty of the other visual responses I had recorded previously, a sense of movement occurred. The dream re-captured my visual response to a narrative that Bill had presented as "one last story" that he "couldn't resist" telling me. The narrative concerned a "happy moment" when he forged a friendship with a highly intelligent and much admired man at university and "felt accepted".

> I can't resist telling you one last story. In (university) I went around quite a bit with (male friend). He was intellectually brilliant, a considerable linguist and the life and soul of any company he ever got into. It surprised me that he wanted to spend time with me. One day in a pub he shyly licked the Guinness froth from his ginger beard, looked around nervously and said, "You know Bill. I am afraid of people." I was startled. For the first time I began to think of myself as his friend. That was a happy ending.

Bill gave the story a happy ending, but the strong visual cinematic response I had to this story suggested that this was an attempt at a form of closure to end discussion about something that was far more risky. I began to tune into the mise-en-scene of the image, evoking Alfred Lorenzer's (1986) notion of the "scenic" qualities of the dream "text". I scanned the dream in the same way a spectator would in the cinema, guided by the camera and editing. I realised from Bill's point-of-view-shot (a technical term from cinematography) that I sensed desire on his part towards this man. It was the kind of editing and camera work that might accompany the start of a romance. I was reminded of the ways in which romantic relationships start to develop in films and the way the camera treats an object of desire for the narrative protagonist. In this case, it was the view of the man's tongue licking the froth from the beer, seen from Bill's perspective, which preceded the narrative revelation about the fear of people. Then there was a cut to a shot where the camera shows the spectator, the slight response of excitement on Bill's face, and the audience senses a budding relationship which, at this point, the spectator cannot know if it is sexual or platonic. I wrote in my diary that the words beautiful and painterly came to mind. I recorded the fact that I had felt distinctly uncomfortable, rather like a voyeur who had seen something she should not have seen. But, significantly, it was the appearance of a *feeling*, one that acknowledged desire and Bill's sexuality, which marked this as a different response. I noticed also that there was a conflict taking place in my mind, as I wanted to avoid the uncomfortable feeling and to replace this with a discussion of the scene's artistic qualities; I became aware that I was using Bill's words to describe the scene as beautiful and painterly.

It took me a long time to realise that I was doing what Bill had done. I was watching a scene intellectually, like a piece of art that helped me to avoid the difficult desirous feelings that had been generated. After living with this feeling for a few days, I found myself wanting to return to the data. There was something about the camera work in the dream that led me to one particular extract from the interview. This concerned one of Bill's favourite films, *Peeping Tom*:

> This film breaks the fantasy (I cannot remember how or when) to make at least male members of the audience aware they have been enjoying some of the voyeuristic aspects of cinema, while the theme is portraying how dangerous and destructive such pleasures become in the protagonist. In the comic counterpoint, the unattractiveness of the older gentleman buying his girlie magazine, also make the male viewer uncomfortable.

I had previously found myself unable to connect emotionally with this kind of account, which I perceived had merely reproduced established aspects of film theory, (see Mulvey, 1975/1988). However, I found an affective connection with the dangers and pleasures of voyeurism and the discomfort of the male viewer that was similar to my own response to my dream. I was able to *feel* some of the danger and risk that had so perplexed me. I realised I had been unable to recognise this because of my own defensive reaction to anxiety. I recognised that I had been adopting processes of intellectualisation towards the material, wanting to produce a particular type of acceptable data.

On returning to the data I found that I was no longer anxious and was not resistant to the idea of risk and what I could now see as the conflictual issues of his sexual and gendered identity. I could understand the anxiety associated with Bill's need to negotiate a new later-life identity, a transition which is recognised as generating conflict (Erickson, 1980). As part of this process of "re-writing the self" (Freeman, 1993), an acute form of après-coup comes into play which acknowledges the risks associated with desire (Laplanche, 1992, 1999) which Bill recognised. I refer back to the last two statements of the narrative about the boys and the bushes: "I shall try and desist in trying to analyse why this might be. Perhaps this is your job?" I might actually agree to his request and analyse what he seemed to be both consciously and unconsciously communicating to me, and therein rests the risk in Bill's dangerous project of "knowing thyself".

Thinking and feeling

I want to proceed, then, with a discussion of affective responses to research texts which develops ideas that were introduced in Chapter 2. Since Lorenzer (1986) promoted the use of psychoanalysis to understand the relationship between society, culture and the unconscious, his ideas have been taken up by the depth-hermeneutic approach to cultural analysis. He argues that cultural texts "provoke the reader subjectively". The filmic responses I have detailed evoke Lorenzer's notion of the "scenic", which is usefully conceived as a filmic or theatrical setting containing actors, action and meanings (Froggett and Hollway, 2010). It is the researcher's task, therefore, to attend to this process of provocation, through a process of emotional attunement to the scene. This involves becoming alert to the range of affective and sensory responses one encounters when confronted with a text. In my own case, I was provoked, intellectually and aesthetically, by the filmic and scenic quality of the data, as was Bill.

I have argued so far that affect plays a role in communication, thought and knowledge, noting that Bion, for instance, begins from the premise that "affects constitute facts and emotional experience is a basis for knowledge" (Bion in Sandler, 2005, pp. 284–291). Affects are knowable and can be unconsciously understood by researchers through listening and observation, even when specific affects or emotions have not been *verbally* articulated. Likewise, interviewees may experience the researcher's thoughts and feelings without linguistic communication (Spezzano, 1993, p. 53) through projective and sensory processes between people, as seen in Bion's (1962) alpha-function discussed in previous chapters. My experience with Bill is an example of an event where thought and affect were constrained, suggesting that attunement was hampered. I would suggest here that my inability to attune to the scenic quality of Bill's data can be understood as an example of the relational dynamics of the defended subject *and* of the defended researcher. This suggests an intersubjective appreciation of the to and fro traffic between the research dyad. In particular, I will draw attention to the defensive use of intellectualisation and how this restricted affective knowing. The use of intellectualisation as a defence represents a novel contribution to existing film and media theory which has largely understood intellectual appreciation as a form of socio-cultural capital (Bourdieu, 1984). For both Bill and I intellectualisation was a defence mechanism which we were using in our specific ways to manage our anxiety.

For Bion raw sense data (what we see, hear, smell, as well as sensations and feelings) are turned into elements that provide the psyche with material for dreams and other unconscious work (as well as conscious thoughts). As I have noted previously, it is only when raw sense data are "linked" that these alpha-elements become meaningful. It is the coming together of sensory data elements that allows thinking to occur (Bion, 1962, pp. 43–47). My incapacity to think, then, can be related to my inability to feel, and affect was the "missing link" in my interpretation of Bill's data.

In Chapter 2, I proposed that interview situations, with their demands to produce knowledge that fits into established disciplinary boundaries, under economic pressures and time constraints, make the interviews frustrating and predispose the research situation to omnipotent thought. For knowledge to be produced, however, frustration must be overcome. On the part of the researcher this means they must relinquish the desire to possess and control knowledge and learn to tolerate doubt (Bion, 1962, p. 36) which often works in opposition to data production for academic publication. This also requires a form of *reverie*, a state of being open to musing, dreaming and sensory data, which is an essential part of alpha-function; *without feeling we cannot think* (Bion, 1962; Ogden, 1999).

I have argued that defensive processes of intellectualisation were circulating between Bill and I. Intellectualisation, as a defence, is characterised by explanations that are "logically consistent or ethically acceptable" (Laplanche and Pontalis, 1973, p. 225), this is the basis on which much qualitative research rests. Bill's interview was replete with statements that could be described as intellectual, both ethically and politically sound. Researchers also need to yield data which enables them to produce this type of account for it to be credible. Bill's data contain examples where this defence is used to manage risk. For example, he recognises the risk in his identification with the old man in *Peeping Tom* and he takes control of the risk through an academic analysis of the film which enables him to distance himself from uncomfortable thoughts and feelings. According to Butler and Strupp (1991, p. 10), when intellectualisation is being used defensively, it is marked by a lack of affect. This defence can also be used projectively. Responses to what is said can unwittingly start to mirror in content and style those of the subject who is using intellectualisation defensively. This could, therefore, account for my desire to interpret my visual responses in the language of aesthetic criticism. The tentative requests for interpretation provoked anxiety for both of us. For Bill, the possible risks were that I might do precisely what he feared and interpret something risky, whereas for me it was the challenge of trying to contain the risk and produce usable data, acceptable to the disciplines within which I work.

In Chapter 2 I explained how knowledge possession can be used defensively to avoid pain and frustration, and in instances of minus K (–K) meaning and emotion are actively denuded of vitality and sense so that discovery becomes impossible (Sandler, 2005, pp. 381–384). The work between Bill and I was marked by defensiveness and intellectualisation on both sides. Intellectualisation is a form of –K which involves a form of knowledge possession, knowing about (an intellectual process), rather than learning from emotional experience and connection with an object. It is also a sign of processes of omnipotence which preclude affective connection between the research parties. This is particularly relevant for qualitative research with human subjects. Bill clearly knows a lot about films, but the risk emerged when he began a process, precipitated by my request for an interview, involving coming to know the films and himself emotionally.

For both parties, there were tensions which predisposed us to defensive processes. I was risking a new method and interpretive approach which might not be acceptable to the disciplines I worked within. I was unable to take control of the data production in a way that appeased my anxiety. Equally, Bill was ambivalent and hence defensive about his request to explore the risks associated with his identity and his viewing tastes. The lack of sensory affective connection was actually beginning to close down my thinking. However, once I had relinquished my desire to control the data, then unconscious anxiety was reduced. Kleinian epistemology would suggest that this represents a move from the paranoid-schizoid position to the depressive position. Froggett and Hollway (2010, p. 189) suggest the latter allows the object to "reveal itself rather than trying to impose preconceptions on it". The fact that I was able to dream about the research suggested that I was able to convert my sensory experience into alpha-elements forming the basis of thought (Bion, 1962, p. 16).

One important theme of this book is that psychoanalysis, as unruly and hence as perturbing as it may seem, posits the idea that it is essential for researchers to be able to draw on their emotional as well as their cognitive experience to understand interview texts. Clinicians, as a matter of course, are supported in the emotional aspects of their work through regular supervision. In this case, I was fortunate to have sensitive and theoretically informed colleagues on whose expertise I could draw. As this adoption of psychoanalytically informed methods such as FANI continues to grow, particularly for PhD research, questions of supervision are important.

Hollway and Jefferson (2010) present a summary of the recent developments in this area. They describe the kinds of supervision some projects have taken up, which varies from infant observation methods[3] to team and group supervision and support. Funded research projects are in a position to buy in professional services and expertise, yet postgraduate researchers are often isolated, as they work alone and on limited funds and may have restricted options. This raises questions about whether it is only possible to work with recognised experienced supervisors and at established centres for psychosocial studies. At the same time, this could also be seen as a potential constraint and could discourage independent researchers adopting psychoanalytically informed methods. I would argue that the way forward concerns the propagation of methodological training in the use of psychoanalysis in empirical work. Those working in the field must press for postgraduate research training and research methods certificates and degrees to include psychoanalytic approaches, which can offer guidance on working subjectively and with emotional experience.

How should we assess the FANI method in the light of this case? The interview generated a question about risk, which I used to highlight aspects of current debates about unconscious anxiety and the defended researcher or subject. Risk was produced out of the circulation of the conflictual unconscious processes of both parties. However, this did not work against the production of knowledge. Though psychoanalysis is often criticised for being used as an interpretive template imposed on the data, as the discussion above and in Chapter 2 shows, adherence to particular disciplinary and political aims can obfuscate data interpretation, and this is not unique to psychoanalysis. In this case, the concepts that I eventually used to help me explain the research process were generated in response to the data. The interview opened with the idea of risk which pointed to the idea of anxiety and conflict. The concept of the defended subject/researcher opened up useful avenues for thinking through what this idea of risk might be, therefore serving an important generative and explanatory function.

Clearly Bill was keen for me to analyse him; but this request is unusual as most interviewees do not request to see the interpretation the researcher makes in relation to their data. The ethics of interpretation have been much debated, with particular reference to questions of power and the appropriate level of involvement of the interviewee. It is seen by some researchers as good practice to offer interviewees a chance to read their interpretation of the interview or to be involved in its construction. There are a variety of reasons why this might prove difficult. I draw attention to the lapse of time between the finished interpretation and the problems of going back to an interviewee after such a time lapse and/or other ethical reasons, taken on a case-by-case approach. This can also be problematic as the nature of analytical interpretation means that researchers often use theories, concepts and what will appear as esoteric language that the interviewee cannot be expected to understand. For this reason, I do not think this is always appropriate. The research project described in this book had the advantage of utilising participants who enthusiastically volunteered to be interviewed. Whilst Bill had a clear agenda of his own, all the interviewees expressed some fascination and puzzlement with their tastes and memories and enjoyed talking about viewing. As Hollway and Jefferson (2010, p. 7) note:

> FANI allows interviewees to follow the threads of their emotional experience as they transform it into freshly discovered meaning... Its aim is to enable interviewees to give answers that reflect their own concerns even when these are not immediately consciously accessible.

This is part of the process of unconscious meaning making which is necessary for emotional well-being (Chodorow, 1999). This particular case illustrated some individual and class-specific defensive processes that might come into play in relation to cinema spectatorship and in response to the anxiety generated by changes in identity in the life cycle. The FANI method enabled me to work productively with my subjective responses even if this was a difficult learning process. Ultimately, this approach helped me to overcome omnipotent defences which work against knowledge production and, as I have argued, these processes are not unique to psychoanalytic empirical research. Once I surrendered emotionally to the dream communication, I was able to begin learning from experience and was able to reap the full benefits of the creative approach to data analysis that psychoanalysis is able to offer the empirical researcher.

Conclusion: Viewing Is Psychosocial

Favourites, objects and personalised viewing

To bring this book to a close I will review some of the key findings of the research and their implications for audience and psychosocial studies. I propose to do this by reviewing what is compelling about the nature of the subjectivities of the viewers studied and their use of texts and viewing. Whilst each case study is marked by its distinctiveness, it is through this uniqueness that it is has been possible to understand why viewing films and television programmes is a subjectively significant experience, and I propose this contributes towards the creation of a psychosocial model for audience studies.

The most important conclusion, as captured by the chapter heading, is that viewing is a psychosocial activity *and* audiences are best understood through interdisciplinary approaches which combine insights from social science, film, television and communication studies and psychoanalysis. The latter is crucial, as it involves the existence of a dynamic unconscious which has its own effects. This premise brings something new to a series of disciplines that have shared similar objects of study but have lacked dialogue. A psychosocial approach overcomes some of the divisions between the study of texts, readers, uses and spectators found respectively in Film, Television, Cultural and Media Studies. The case studies demonstrated intricate viewing configurations, emerging from a range of mutually constitutive, social, emotional, historical, discursive and unconscious processes that can only be fully understood using a range of approaches and methods. Whilst the case studies utilised a range of psychoanalytic ideas, the overarching body of theory used here, the most effective for audience studies, has been Object Relations This is in part due to its inherent recognition of contexts, which involves the social and agency, something most audience researchers would see as essential. It was the universalising tendencies and perceived lack of recognition of social relations outside the Oedipus complex that caused so much suspicion and resistance to earlier attempts to wed psychoanalysis with audience studies. The idea that humans' emotional relationships with objects, both human and non-human, is at the heart of self-existence makes it all the more urgent that researchers understand the multifaceted experiences of engagement with the media. Viewing is both a cognitive and affective experience. Films and entertainment texts are designed to affectively and emotionally move their audience as part of their "use value" or character. Media texts and technologies exist for viewers as evocative objects (Bollas, 1992, 1995) which may or may not have transformative qualities. The subject can use media texts as objects to elaborate and bring the self to life in the Winnicottian sense of the spontaneous true self. This happens through a range of projective and introjective processes, and encounters with objects are subject to the work of unconscious phantasy in the creation of meaning. Conversely, particular texts, their use and associated viewing practices might also be born of defensive processes and responses to biographical events.

To understand and explore these ideas, I used the mundane, but at the same time generatively rich, category of favourites. Favourites are impact texts, in that they leave an impression. This can be in the form of a memory or emotional response that makes them unforgettable as well as being texts that are returned to and revisited for a variety of reasons. Favourites can be linked very closely to what has been and must remain a significant category in audience research: identification. Identification is important psychoanalytically because it points not only to a cognitive recognition of some similarity with a quality or trait as a feature of an object (which can also be a moment of narrative or plot development in a text), but it also points to an emotional connection with an object. This connection can then lead to it being put to particular uses, and this can be seen as one of the features of the relationship the audience builds with the text/object. It is this combination of favourites and identification that explains viewing as both a pleasurable experience and one that is often uncomfortable and anxiety provoking. For example, as we saw in the case studies, it was Chris's identification with *Barton Fink* as a weak man, failing professionally, that made him feel so uncomfortable and Sophie's identification with an abandoned doll under a teenager's bed which chimed so clearly with her own sense of loss.

Whilst viewing and favourites can be explored within the category of object use, using objects to bring the self to life is not always pleasurable or enjoyable. As the case studies demonstrated, objects can be used to address conflict. Viewing is an activity that can generate different forms of anxiety, but this anxiety is always linked to historically and socially specific events, processes and discourses of the outer world within which the viewer is interwoven. Audiences are also resilient and develop and adopt unique ways of managing different types of anxiety with varying degrees of success, as exemplified by Bill's intellectualisation, Daniel's use of television to fill up time, and Mary's creation of an internal space where her real thoughts and feelings are kept safe and preserved from the disapproval of her mother. It was these insights that led me back to uses and gratifications models. Despite their current unpopularity in Media Studies, the idea of television being put to use to meet needs is well worth retaining and revisiting, as it can provide a window on the relationships between inner and outer worlds when conscious and unconscious processes are acknowledged.

The use of media texts for psychic and social needs is aided by new media technologies which enable viewers to *personalise* their viewing. This can also be linked to other processes of individualisation as the mark of 21st century life. The study has indicated, however, that the changes associated with individualisation, reflexive modernity and other developments such as social mobility, make class and gender anxieties all the more acute and prominent, as traditional forms of group belonging are under threat due to economic and social change. The case studies have shown that the experience of social belonging can be understood, in part, through viewing that includes gendered understandings and interpretation of biographical experiences.

Family viewing, identification, personalisation

As Television and Media Studies continue to grapple with the pace and impact of technological changes, there is a need to reassess the direction of audience studies, the shift from viewer to "viewser" (Caldwell, 2003) being one example. Family viewing, which was typically associated with the possession of only one television set and limited programme choice and has been such a significant part of television theory and research in the past, has also changed. These changes have emerged because of a combination of social and technological developments. In contemporary society, more people live, and hence view, alone in domestic settings, often some distance from the geographical location in which they grew up. They also view in different spaces and on new kinds of technologies other than televisions, such as computers and i-Pads. Despite this, as this book argues, important figures from the viewers' pasts, such as parents and siblings, are present unconsciously and imaginatively when they view, without the need to be physically present, and this happens regardless of the technology. Thus, it is important to hold on to a notion of family viewing that recognises that families, and the traces of their earliest experiences (which are subject to processes of unconscious phantasy), are present with viewers as they watch television and films, even if they are not conscious of this. This is further enhanced and intensified in the context of personalised viewing which creates new emotional possibilities and tensions for viewers.

Theories of narrative were traditionally based on the importance of the (male) hero and his quest which influences the audience's identification (Mulvey, 1975; Todorov, 1977). Narrative theory proposes that stories begin with an equilibrium which is disturbed early in the film to be replaced by a new balanced state at the end of the narrative. One of the characteristics of a favourite text is that it contains an aleatory factor, a moment of surprise that creates a form of narrative disturbance which halts and disrupts the narrative for biographical reasons, even if, initially, the viewer is not fully conscious or aware of the reasons for this. This can be at any point in the film/television text. This can mean that the "climax" of the narrative occurs, for some viewers, at a point other than one found in a dominant reading of the narrative. I have therefore suggested, following O'Shaughnessy (1994), that a new dimension be added to theories of reading texts; one that considers the emotional and biographical forms of decoding. Established approaches, which examine the form and meaning of texts to suggest preferred or the most obvious dominant readings, remain useful starting points from which to work. We also need to understand, however, why a particular reading "makes sense" (I prefer this term to O'Shaughnessy's original conception of a "mis-reading") for particular viewers. Thus, in Sophie's case for example, it is possible, by taking account of her biography, to understand more fully her complex and seemingly excessive responses to certain texts (such as the scene of the lost ball in *Castaway*) and her desire to engage in "dis-pleasurable" viewing.

Another feature of aleatory interpellation is that the initial element of surprise can also signal a text or moment as something evocative and useful for emotional or transformational work; for example, when Chris sat through an episode of *Frasier* and realised it was "really good." This later became a very important text around which transitional practices to appease anxiety were built. Aleatory film *memories* are also important, and I used the concept of après-coup or afterwardsness in a pragmatic way, to show texts can be returned to in the light of other and current experiences. Sophie remembered *The Wizard of Oz* from childhood and went back to it when she realised she identified with Dorothy and her journey because of her experiences of using friendships in coping and surviving loss. The mantra "there's no place like home" became woven symbolically into an extensive and complicated project of object use, to create a haven of security and stability (her flat full of "stuff") of which Sophie was in control. As Sue's and Mary's case studies also showed, aleatory film moments (*Father of the Bride* and *Big*), become significant memories because they present the viewer with scenarios which generate powerful intergenerational identifications and evoke events with biographical significance.

Psychoanalytic perspectives on fantasy have called into question the idea that there is always a necessary correspondence in gendered identification between viewers and on-screen characters, as identification can be fluid and move around characters and objects within a scenario. The study suggests that other kinds of multiple and simultaneous identifications, such as intergenerational and cross-gendered, can explain the viewer's identification with on-screen characters, plot and mise-en-scene. The study has shown that biography and experience must be considered if we are to fully understand these particular types of identification (see Chapters 3–6). In turn, the case studies have shown that in audience research not enough is understood about the role of internal objects on identification and the creation of meaning, which appears particularly relevant in the case of favourites. Furthermore, viewing in domestic settings (or what viewers might call "home") is a highly personalised space, containing a variety of objects with symbolic value to the viewer (see Sophie, Chapter 4), which as I have argued gives the home viewing space a particular intensity in which memories and primitive processes, associated with families and watching television with families, can be evoked.

Whilst these processes might be powerful, they are, at the same time also everyday processes, such as the use of television to create a holding environment. These quotidian domestic processes are distinctively different in quality from intense moments in the cinema that might occasionally cause spectators to shed a tear. It is difficult to put cinema to work as an activity in the same way that viewers can at home. In fact, these processes of personalisation that new technologies offer could be said to provide ever *increasing* opportunities for psychic engagement and investment.

Families affect viewing and identification in different ways. At the same time, examining families and identification illustrates how social processes are experienced and given meaning through significant relationships within them. This was seen, for example, in the case of Chris who, as a result of social mobility, experienced conflict and loss because of the changes this brought to his relationship with his working-class father. For Sue, the experience of a being a working-class woman in the 1970s and 1980s was consciously and unconsciously explored through her complex identifications with her younger male sibling, interwoven with Christopher Eccleston as a significant and evocative object. These tensions and conflicts can be understood through examining these processes of identification (and dis-identification) with texts that become favourites. This brings me to another key finding about the way in which families have an effect on identification and viewing: siblings.

Siblings

One of the aims of the book is to reflect on psychoanalytic concepts and theories, mapping change within both areas and assessing their appropriateness for use in the particular space of audience research. The findings so far have been shown to have relevance for both audience studies *and* psychoanalysis, particularly in the under-researched area of siblings or, more specifically, "the mental representations of childhood siblings as influential object representations or identifications" in the adult's internal worlds (Volkan and Ast, 2014, p. xi). The impact of siblings on the inner world of children has been explored in psychoanalytic literature, but much less is known about the impact of sibling experiences going into adult life (ibid.). The case studies suggest that the role of siblings is best understood psychosocially rather than intra-psychically. The study pointed to siblings being responsible for a range of powerful responses which impact of identification with texts. There are texts and viewing memories that point to the on-going significance of sibling rivalry and hostility (see Chris and Mary, Chapters 3 and 5), and texts which point to the important role siblings play in processes of separation and individuation (Chapter 5), and the existence of Oedipal sibling triangles which make Sue's case study so fascinating. At different stages in life, siblings can be associated with different forms of anxiety as well as ambivalent and desirous identifications, seen for example in Sue's worry about siblings having a relationship in *Jude* and her overlapping identifications between her brother, son and Christopher Eccleston. Chapter 5 also demonstrated how siblings were associated with common unconscious phantasies concerning the primal scene, incest and family romance (*The Birds*, *The Singing Ringing Tree*).

Clearly, siblings matter. The arrival of siblings confronts the child with adult sexuality and a plethora of fears about being displaced and lost. What is interesting is the way in which film memories point to the way texts provide imaginative raw material that can also be subject to projections and phantasies that emerge from the inner world of the child. Film images can become the visual objects of screen memories (*The Birds*) or symbolic representations of childhood fears and repositories of aggression (*The Singing Ringing Tree, Hush Hush Sweet Charlotte*). Most obviously, this shows how the effects of films on children are, in part, rooted in existing fears and experiences that the child contends with in its life. The text is only a piece of the jigsaw, and it is the rich sensory potential of film, which film theorists have noted, that makes it particularly open to the work of alpha-function. The film moments in *The Birds* and *Hush Hush Sweet Charlotte* operated as linking elements, but this was only possible when coupled with the settings of home and the presence of family members who would also have been communicating consciously and unconsciously with the child. This shows the uniqueness of the home viewing situation. It is impossible that this combination of factors, which resulted in these films being perceived as frightening, could occur similarly in the cinema. Of course, the home is also a risky setting, in that television viewing is never subject to the strict censorship and classification guidelines cinema must abide by. Both of the films that frightened Sue and Mary so much were aimed at adults and were shown after the watershed, but children can always by-pass these restrictions. Even so, the combination of life events, the home and the conflicts of the child's inner world, made *The Singing Ringing Tree* fairy story, shown as an after-school programme, terrifying. For Mary, this was linked to sibling rivalry and Oedipal conflict, captured symbolically in fears around animals and birds (*The Birds* and *The Singing Ringing Tree*). As we have seen, exposure to frightening films can occur in the home, and the resultant fear is a combination of lived experience and the perceived content of the text, which is understood through the lens of the inner world. Here, errors made around cousins and siblings and reasons for animal attacks merge with other aspects of lived experience to form new and frightening meanings. The memories remain powerful for the women because they return in this memory to the child who was struggling to understand a number of momentous events. This also explains why they "don't know" the reason for the memory remaining so powerful and "terrifying".

In adult life, sibling triangles and allegiances can form where sibling rivalry and hostility is reactivated in response to life events and Oedipal anxiety and can help to explain some viewers' complex identifications capturing both love and hostility (See Sue Chapter 5, *Our Friends in the North*, *Hearts and Minds*, *Jude*; Chris, *Frasier*, *Some Like It Hot*, Chapter 3). Siblings in psychoanalysis have also been shown to play a role as "actors and actresses" in unconscious phantasy where there is a "story-like" quality but its meaning may be latent for the individual (Volkan and Ast, 2014). This is why narrative films and television are so rich for providing source material for viewers to take up and work on in the inner world, as well as to project bits of the self into, and explains to some extent why Christopher Eccleston, for instance, has become such an important star for Sue. The parts he has played chime with the intergenerational intensity and angst of her brother and eldest son.

Using viewing objects to bring the self to life is only one aspect of object use and one feature of the relationship viewers develop with texts and viewing. The case studies show that viewers do an immense amount of work on the difficulties and conflicts that also form part of experience. I have followed Layton's (2004, p. 48) position that unconscious conflict results from contingent historically specific norms, as well as universal laws such as the Oedipus complex. This account points to conflicts resulting from the demands of normative discourses in hierarchical societies and how they are taken up and played out in specific families. We saw in Mary's case how the demands to be traditionally feminine were unique to her specific family. Femininity became something she was antagonistic to as a reaction to her siblings. This was further complicated by her position in the family and linked to intergenerational traumas she carried for her parents, particularly her father who had experienced being judged and not good enough. This can be understood through the shift from an idealised father (*Father of the Bride*) to a more ambivalent discussion of a less than ideal family in *Meet the Fockers*.

Bourdieu (1984, 1987) has argued that processes of socialisation resulting in the transmission of the habitus are passed unconsciously and intergenerationally (see also Reiss, 1981). The case studies have shown that this is not a conflict-free process, and Mary's study pointed to the intergenerational transmission of trauma. These observations follow Layton (2004, 2002b, see Chapters 2 and 5) who stresses that the need for approval in families is one of the ways social hierarchies, as systems of value, are maintained and that this is the result of bourgeois ideology. So, for Mary, there were demands to perform femininity as defined by dominant ideology and to be socially mobile, through marriage to a middle-class professional. This, however, repeated her father's own trauma, of being the figure that was not good enough for her grandparents. Mary resisted the demands to conform, through her vibrant inner life where she could be a bad girl and masculine, but also through her defiance in marrying a working-class man. Mary appears to have put a stop to this trauma being transmitted further in her choice to have only one child following the birth of her son, who she (unsurprisingly) admits can "do no wrong". This shows not psychological inertia, but her resilience and ability to overcome these intergenerational difficulties (*Pretty Woman, Tenko, Meet the Fockers*).

Repetition

Relationships with texts are often built through processes of repetition, including returning to a text and other associated repetitive viewing practices (I have extended the idea of repetition to also include returning to "unforgettable" viewing memories). I have proposed that a psychoanalytic understanding of repetition adds a new dimension to existing audience research. The intense relationships with texts are often the result of technological innovations and also greater opportunities to consume. In Daniel and Sophie's studies, the ability to select, purchase and control viewing became a marker of independence. It also involved putting the text to work emotionally, which concerned the text and its content, certain practices the viewer connects to the text (such as block viewing or watching whilst taking sick time off work) and the location of the text within wider patterns of object use (*Twin Peaks, The Wizard of Oz*). The idea of repetition, as it might be understood in the purely psychoanalytic sense as something that involves mourning, (or resistance to mourning), through replaying the loss or trauma over and over, was shown to be something that is relevant for audience studies. The studies have shown that it is possible to view some favourite texts as repetition scenarios, but in the psychosocial sense of conflict that emerges out of contingent historical circumstances, which form the foundations of a repetition scenario. This was seen most clearly in Sophie's favourite films, which I have called mourning texts, which are also associated with generating signal anxiety associated with abandonment: the loss of *Wilson* in *Castaway* and *Jessie* abandoned under the bed in *Toy Story II*.

Sophie's complex object use and consumption were produced out of loss but interpreted against a historically specific set of circumstances and relations in her own family which have given it a particular shape and form. She is only too aware that when she watches *Toy Story II* and

Castaway she is choosing to watch films that make her cry and feel "bad", but the reason for this is unclear to her. I have suggested that the scenes capture and replay traumatic experiences from her childhood. Returning to the text is similar to returning to a trauma and replaying it over and over. The trauma of abandonment for Sophie is experienced in the light of processes associated with new right conservatism. Sophie's emotional security must be set against the backdrop of the rise in consumerism associated with the 1980s. She experienced emotional loss, in material terms. Following her father's departure the family lost his income, their status, when they moved from a comfortable home into a council property, their possessions and Sophie's much-loved dog. It is not surprising that the possession of things should be so appealing to Sophie, as they signify safety and security as well as status. This shows the relevance of incorporating a historical perspective.

As noted in Chapters 2 and 4, repetition compulsions are not just the wish to return to the trauma to avoid mourning. They are not static but are motivated by the wish to find a way out of the trauma and bring about change, however minute. This is something Langan (2014) takes up in citing Ferenczi's idea that trauma is not a unique experience but something that is part of the fabric of ordinary life, which in turn awakens the mind to finding an escape route. I am pointing to the use of films and television viewing as one such avenue. This is one of the ways in which, in Winnicott's terms, we are able to "go on being", such as Daniel on the sofa in a darkened room watching *Twin Peaks*. In Sophie's case, the texts she chooses to watch make her feel "bad", but she is drawn to them in the hope that she can initiate some changes, so that she can ultimately feel *better*. Sophie's case illustrated favourite texts as repetition scenarios (*Toy Story II* and *Castaway*) because they were specifically concerned with loss and mourning and also shows how the content of texts might generate forms of signal anxiety concerned with helplessness. This kind of repetition is different to Chris's use of *Frasier*, which has anxiety placating functions because the text has the ability to *resolve* conflicts through the narratives and humour for this specific viewer, making this a transitional text and viewing experience.

Object use and taste

Objects only come "alive" in their use (Bollas, 1992, 1988), and this study has shown that audiences put television and film objects to use through a range of conscious and unconscious processes. Whilst the book did not set out specifically to explore the nature of television viewing, programmes and films as objects, clearly they are complex objects which provide rich imaginative pickings which audiences can and do put to use. Likewise, the study has concentrated on the specific experiences of audiences. From this, it has also been possible, in a limited way, to suggest reasons for the appeal of particular genres and texts, such as soap operas (Chapter 2), time-travelling and time-based texts (*Back to the Future*, *Big*, *The Time Traveller's Wife*, *Forrest Gump*, *Dr Who* and *Simon Birch* (Chapters 3–5) and in support of existing accounts of auteur texts for example David Lynch and *Twin Peaks* (Chapter 3). I have noted the significance of texts, time travelling texts being good examples, which offer possibilities textually for intergenerational identification. Such texts are evocative of processes, where in phantasy, the viewer can go backwards to perform reparative work, address Oedipal conflicts and make changes to their life. This kind of text might allow the viewer to borrow the narratively constructed structure of feeling of an era (such as the 1950s for Sophie), and in the spirit of this time travelling bring it back to the future and incorporate this imaginatively into daily life.

It has been important to draw attention to the emotional and psychic resourcefulness which these viewers have shown, but I have also been keen to stress that the viewers' relationships with text are not always celebratory or resistant. Object use can also be defensive and develop into conservative patterns of usage. This is because ordinary life is difficult, even traumatic (Ferenczi in Langan, 2014, p. 22), and this must be recognised if we are interested in thinking about the emotional costs of social change. Processes of personalisation are contradictory. Daniel Miller's (2008) argument is that people, who have a lot of objects which they have strong relationships with, will have closer relationships with people: the more objects a person has, the more connected they are to other people. He argues this without recourse to anything psychological. The case studies have shown this is too narrow a conception.

It has been argued that our need to endow and tint the world with our idiom involves important unconscious processes that are associated with emotional well-being. However, the demands to produce a socially acceptable identity and habitus can also inhibit the spontaneity and creativity that are important components of these personalising and idiomatic processes and lead to forms of conservative object use. As we have seen for Sophie, her excessive consumption and inability to separate herself from "stuff" also creates problems for her. She hates her job, but is unable to change it as she is economically locked into credit card bills, linked in turn to her need to continually consume, on which she has made her emotional stability also rest. This is something Miller's (op. cit.) argument captures, as she is also using her relationships with non-human objects to mend her relationships with human objects. My argument is that this invariably fails. Nor is a close attachment to things indicative of the quality of other relationships as Miller argues, it is possible to have a lot of things, but they can be part of a "narrow colony of objects" (Bollas, 1995), indicative of conservative object use.

Understanding Sophie's attachment to objects in films such as *Castaway* has also thrown light on the power of marketing and advertisement strategies in contemporary capitalism to present consumer goods as evocative objects with the power to transform and to perform emotional work. Ultimately, the consumer realises the object cannot fulfil this promise and, like Sophie, moves onto the next object.

For Daniel, new viewing technologies are functional for anxiety management as he uses them to facilitate a holding environment. Television viewing can also be understood through the idea of the autistic-contiguous position (Ogden, 1989, 2004), which explains the appeal of lying on the sofa in the dark, barely moving, as a primitive defence mechanism. This helps him manage the anxiety generated by time but, paradoxically, this activity is also contributing to increasing social isolation. Chris's desire to "slob out" is over-ridden by the anxiety generated around his professional identity. He spends a lot of time feeling frustrated, whilst desiring the kind of relaxed autistic-contiguous viewing that Daniel is able to give himself permission to enjoy. He also recognises that culturally a negative value is placed on this kind of television viewing, (hence his use of the negative term "slobbing out" rather than seeing this as relaxation). He attempts to manage his anxiety by collecting, cataloguing and working intellectually with texts which, in turn, generate more anxiety. This, once again, reinforces the intensity and the psychic complexity of the viewing experience in the home and that object use is not always generative of growth.

Taste is also best understood psychosocially, involving class and gender socialisation that is shaped by dominant ideology, which is interpreted in specific ways by families. The case studies show that the struggle to adopt a habitus and develop a disposition, so that one might be recognised as a member of a particular class, seen most clearly in Chris's study, is fraught with difficulty. Discourses around the value of popular culture, which position television viewing negatively, contribute to the tensions expressed about television viewing. This idea that watching TV is "not good" for people was expressed to some extent by all the interviewees, but as we saw this was particularly acute for Chris, Daniel and Bill. The anxiety around negative discourses about television and taste also suggests possible links to the demands to perform particular types of autonomous masculinity. Additionally, those most embedded in the middle class are not conflict free, but adopt the skills and training this provides to manage this conflict, such as intellectualisation and self-reflexivity, which is also a feature of post-industrial modernity (Giddens, 1991; Beck, 1992).

Bourdieu's (1984) theory of cultural capital explains why taste is important and how it relates to hierarchical differentiation and translates into forms of cultural and economic power. He recognises that, for the habitus to work effectively, it must become unconscious without fully acknowledging *how* these unconscious processes work. Having stated this, in the case of taste, I have stressed that an individual's emotional agenda affects choice. Selections are never transparent as they contain aspects of early life experiences and families play a significant and on-going role in this, particularly where favourites and very personal texts are concerned. Thus the habitus might never become second nature and is lived as a painful and conflict-filled experience, suggesting that, where social mobility is concerned, it is not always easy to develop and feel comfortable with a new habitus and disposition (Chris, Chapter 3; Bill, Chapter 6). Conflicts occur in response to the demands of established cultural values and hierarchies and also to changed social circumstances, which create tensions in relationships with significant figures as Chris and Bill's studies showed (Chapters 3 and 6). For this reason, it is important to recognise that early experiences and families play a more complex role than just agents of socialisation in the case of class habitus and taste.

Whilst there may be evidence of increasing processes of individualisation in society, the case studies reinforce the relevance of class and gender in understanding processes of individualisation. Ideologically capitalist societies are marked by the ideology of the free individual, who is self-reliant and constantly working on constructing a successful self through consumption (Layton, 2002b; Ouellette and Hay, 2008). Viewing, as part of displays of taste, creates a range of tensions which link taste not only to the performance of class but also of gender. As argued in Chapters 3 and 6, this can be discussed in terms of binary opposites with one half of the pair having more "value" than the other, in this case masculine versus feminine. Thus, for Chris, anxiety around taste and social class is also worked over in terms of gender, as he continually struggles over his masculinity, generating anxiety as he identifies with weak, feminised males in films (*Some Like It Hot*, *Fletch* and *Barton Fink*). Conversely Bill, adopting the techniques of reflexive modernity eloquently, discusses being "accepted" by both males and females in his life which can be seen as a controlled display of defensively performed middle-class masculinity. In contrast then to Beck (1992, p. 136) who sees contemporary society as "ego-centred" with cultural products being used, not just to display taste and show membership of a group, but to define the individual as an *individual*, the case studies show that there are a range of unconscious conflicts which generate anxiety, which the ego can only superficially manage. Displays of taste and object use can also be linked to the anxieties and effects associated with changes to social circumstances: living alone, divorce and social mobility (see Beck, 1992). Worries about "spending too much time watching TV" are common (Chapter 3). The appeal and popularity of texts involving the replacement of family by friendship groups such as *Friends* and *Beverly Hills 90201*, whilst special to Sophie, have also been linked more widely to these kinds of changes.

I have taken the position that taste, as delineated by Bourdieu and shaped by the demands of bourgeois ideology, is still open to the work of personalisation by subjects. Audiences can use objects and practices to elaborate their character, even if this is a conflictual process. Whilst the work of Bollas has been important in understanding object use, his work is primarily as a clinician and he is not a social theorist, so it is disingenuous to criticise him for not theorising the social. He is interested in experiences, which cannot be divorced from the social, but his purely psychoanalytic approach does have limitations. He lacks a clear perspective on the way the social might impinge consciously and unconsciously on the individual's ability to choose and use an object to elaborate the subject. Whilst he acknowledges forms of conservative object use (Bollas, 1995), he does not adequately theorise how membership of certain groups, such as social class or gender, might generate particular anxieties around acceptance. For instance how systems of social and cultural exchange value generate anxiety which might make it impossible for the subject to put the object to work in anything other than a conservative form. He also bases his theories (see Bollas, 1987, 1988, 1992, 1995) on illustrative examples of generative object use, where the subject uses what Bourdieu would class as legitimate or high culture such as classical music, literature or art, rather than popular culture. Bollas also attributes complexity to the former, without acknowledging popular culture as a more conflictual site because of the differing values they carry discursively. I want to argue that social class and economic reality can sometimes predispose the individual towards the use of only a limited colony of objects that Bollas (1995) describes. Excessive television viewing might be considered as an example of limited object use, as shown in some of the case studies. It is also possible to see there are opportunities for generative and reparative, as well as conservative, object use. Thus, the processes involved in the establishment of taste and favourites are not entirely processes of the ego. As I have argued, they are the result of a combination of inner processes and needs, past experience and the social reality that the individual lives in the everyday.

Method and concept evaluation

Proposing to understand viewing psychosocially meant experimenting with new methods and approaches, most significantly from object relations psychoanalysis, whilst remaining within research traditions that were based on talking to audiences in interviews. Free Association Narrative Interviewing brought something new to audience research and was relevant for a research project on favourites because the method was designed to illicit the ways that subjects attribute meaning to experience and also the emotional significance of texts and viewing practices. The method is also ideal for examining processes of personalisation in viewing which were central to the research project. The narrative and biographical features of the method not only explained the personal significance of a text, or viewing practice, but helped to locate the viewer historically and highlighted the political and economic context in which their experiences take place. The method was able to capture past and present concerns, as seen across the case-studies, where identity is reconfigured (Bill and Chris) or where there are moments of intergenerational or cross-gender identifications (Mary, Sue, Chris and Daniel). The context that was revealed in Mary's narrative about *The Birds* was central to understanding why this event was so frightening. It would have been unlikely that I could have produced data suggesting these meanings if I had used a conventional question-and-answer technique. My approach enabled me to produce greater understanding of the resilience of human beings in response to the difficulties they face without being uncritically celebratory and ascribing positive political motives to this.

However, as Chapters 2 and 6 illustrated, using psychoanalysis is not straightforward, and there are some ethical problems which Frosh and Baraitser (2008) rightly noted. These are not unique to psychoanalytically informed audience research. All interview situations can generate unconscious anxiety and conflict, hence my reason for revisiting Seiter and Borchers' oft cited piece in Chapter 2. Any research topic that is interested in life experience, if we follow Ferenczi's (in Langan,

2014) conception of trauma as part of life, is bound to be difficult for interviewees and interviewers. An important psychoanalytic insight is that there is no way to predict (with any reliability) the extent to which a particular topic might provoke emotion or anxiety. My topic of favourite films, on the surface seems completely benign compared with what might seem more serious topics that social/psychological research may deal with. As we have seen, however, this everyday activity can be anxiety-ridden and embedded in difficult emotional experiences. Despite this, all researchers have in place the established safeguards that they can draw on, such as stopping the interview and anonymising data. The psychoanalytic ideas which inform the method, such as learning from emotional experience, the use of the researcher's self, the role of affect, as well as understanding the role of discourse and language, can help researchers acknowledge and accept the messiness, doubt and uncertainty that are features of research gathering, which can get in the way of the development of genuine empathy and communication. Thus, the FANI method, through its recognition of the presence of anxiety and conflict and the role played by phantasy, provides the basis for a realistic assessment of the processes involved in data production and communication between researcher and interviewee and offers "an enriched, more complex, nuanced and more humane and ethical view of the subject" (Hollway and Jefferson, 2000, p. 155).

The methodological insights represent important contributions to communication studies as well as television and film studies. It is no longer tenable to discount theories of unconscious communication, however complicated they may be to work with. I have also argued for a process of triangulation between text, audience, researcher and other forms of external evidence. This has other implications. It is essential that audience researchers do not reject the text (and theories of reading) or prioritise it, as explained above. Of course, this approach is best suited to research interested in experience and meaning and might not always be as effective for research that looks for more generalisable patterns across large audiences. It is also useful for memory work and oral history in film and television studies.

To this end, the study and its aim of assessing the viability of psychoanalytic concepts outside the clinic was useful for both psychoanalysis and audience research. The findings illustrated the continuing importance of established categories such as identification and demonstrated the complexity of processes associated with being both an active and passive audience. Whilst certain topics had recently become passé, such as family viewing and uses and gratifications, the findings dealt with some of the under-theorised gaps, particularly the idea that researchers cannot know the psychological uses audiences put TV to (see Elliott, 1974) and the idea that families are no longer really important in viewing due to technological developments. In fact families, as internal objects, and the home as a viewing space have taken on new importance precisely because of these technological developments. I hope it has also shown how productive it is to promote dialogue between disciplines and approaches that were previously divided.

Equally, this evaluation of concepts has been useful to psychoanalysis, preventing it from becoming clinically focussed, esoteric and isolated from social science and humanities. The study has added to knowledge about the significance of siblings and pointed to the importance of considering social processes more fully in individual psychoanalysis.

As I have shown, the breadth and depth of psychoanalytic theories and concepts have much to contribute to audience studies whilst remaining true to a discipline that is interested in understanding emotional costs and difficult experiences and their effects. However, if the social sciences and humanities are going to adopt psychoanalytic methods, then the idea of supervision must be considered. It is difficult for single researchers working alone to do this, as I noted in Chapter 6. It is important to work with others and preferably a team who have some experience in this kind of approach. In the clinical setting, supervision is essential, so this poses the question, "Can psychosocial research only be carried out in collaborative projects where the researcher has a chance to explore their responses with others as well as exploring ethical issues?" There is no answer to this but it does suggest that research training needs to consider the best ways to prepare researchers for empirical psychosocial projects. I am optimistic that demands for team and group supervision may contribute towards psychosocial studies leading the way in a return to collaborative work, of the kind that was so productive for cultural studies in the past. This has been lost due to institutional developments associated with research funding in recent years, as a result of policy developments which have introduced a competitive market into higher education research, due to global change.

Conceptual and theoretical research generates as many questions as the research answers, and this is in the nature of coping with the uncertainty of not knowing, which Bion (1962) articulates. The analysis of data is always provisional. I am aware that the material included in this thesis is only one track of many that were possible. I have demonstrated the value of the single case in conceptual research without needing these concepts or findings to be statistically generalisable or replicable.

Psychosocial research is sometimes accused of claiming to know people better than they know themselves (Frosh and Baraitser, 2008). I disagree, as theoretically and conceptually grounded theory starts from the position that there is always more about interviewees than we know. The approach adopted in this research captures a small slice of the complexity of my interviewees and brings to light areas of the viewing experience previously not captured. It shows the vibrancy of the imaginative inner world and the psychic and social resourcefulness of viewers, to carry on being and overcome difficulties. It also demonstrates the intensity and psychic appeal of entertainment texts. In this sense, then, the book has achieved its aim. The everyday activity of viewing television by ordinary people shows that far from ordinary, everyday life and cultural activities are complex, difficult, extraordinary and fascinating things.

Notes

Introduction: Puzzling Viewing

1. Elizabeth Cowie's (1997) work is notable for her use of Freud's ideas on fantasy to bring in a more historically specific understanding of identification desire and cinematic pleasure. Cowie proposed that films offer *multiple* points of entry for identification in films, and she argued there was no automatic correspondence between identifications with characters of the same gender. The spectator's identifications could oscillate between differing characters or even objects and events within one scenario. Cowie's reading of the films *Now Voyager* and *The Reckless Moment* showed how films create possibilities, in this instance, for forbidden female desires which subvert the existing social prohibitions of the day. Her psychoanalytic reading was alert, therefore, to the historically specific conditions of viewing.
2. See also Kuhn (2013) *Little Madnesses: Winnicott, Transitional Phenomenon and Cultural Experience,* London: I.B. Tauris.
3. Within British psycho-social studies there is debate about the concept of transference and its relationship to the clinical setting and outside as used by researchers. See "Special Issue: British Psycho (-) Social Studies" in *Psychoanalysis, Culture and Society,* Vol. 13, No. 4, December 2008.
4. Bennett et al. (2009) recently completed a review of Bourdieu's work with a large- scale survey of the organisation of cultural practices of the British public. Concurring with Bourdieu's (1984) study, they found that systematic patterns of cultural taste and practice which correspond to social class divisions endure. Bourdieu, in his original study, did not pay sufficient attention to television viewing, and television is interesting as it is often thought to be a democratic medium. Bennett et al. found that film and television viewing in particular had weaker patterns of demarcation than other fields such as music and art. They found that watching television and cinema-going are two of the most widely shared cultural practices throughout British society.
5. Each participant was provided with details of the research, with the standard British Psychological Society ethical guidelines observed. Consent forms were supplied and signed. Names and identifiable information about geographical locations and recognisable work places to protect identities were omitted or changed. In Chapter 4 Sophie is a pseudonym; however, I was given specific written permission to use her maiden name in relation to her interview about the film *Castaway.* The sample was recruited through a process of snowballing.

1 Favourites, TV and Home: Psychosocial Perspectives

1. See Neale's (1986) study of Melodrama traditionally associated with "weepies". He notes the way narrative events are motivated and ordered. These films tend to be marked by an excess of affect and effect over cause.

2. Emerging from within psychotherapy in the USA, "Film Therapy" or "Cinematherapy" (Berg-Cross, 1990) has found its way onto the popular "self-help" bookshelves. For example, Peske and West's (2002) *Advanced Cinematherapy: The Girl's Guide to Finding Happiness One Movie at a Time* finds a film to solve every *"Bridget Jones"* man problem and dilemma.
3. Whilst Sutton follows Laplanche's use of the term afterwardsness, the original term is German and comes from Freud: Nachträglichkeit. This was originally translated into English as "deferred action" and only later rendered as afterwardsness.
4. Tania Modleski in her notable psychoanalytically informed study of soap operas *Loving With A Vengeance* (1982) uses repetition compulsion to account for identification with the villainess found in all American soap operas. Her argument is that women tune in everyday to see the villainess, who is continually faced with events that would make most totally helpless, somehow manages to turn these events, if only temporarily, to her advantage. Modleski argues that repetition here concerns fantasies of power about the rejection of feminine passivity and mastery of events. So to some limited extent it is compatible with the way I use the term in this research as a return in part to motivate some kind of change.
5. For a discussion of the differences between the approaches of Winnicott and Klein, see Robert Young's on line paper *Potential Space: Transitional Phenomena* (http://human-nature.com/rmyoung/papers/paper55.html).
6. Silverstone does acknowledge this and is keen to argue this need not be pathological and a negatively "regressive" experience.

2 Psychosocial Methods and Audience Research

1. For detailed account of FANI method and critical evaluation by researchers who designed the method see Hollway and Jefferson (2010).
2. When the child is distressed because it is hungry or is uncomfortable due to the sensation of a soiled nappy, the infant does not experience this hunger or excrement as something separate, but phantasises that the object is an attack. Anxiety is generated in response to fears that the object is actually lodged inside the child and threatens his safety and survival. This experience may be preserved internally, as a bad object, which can attack the child from within. On the other hand, when the child experiences something pleasant he preserves this in phantasy as a good internal object. This explains how the inner world can be populated with a range of good and bad objects.
3. Neuropsychological research has also found evidence to support the idea that relations with objects are a defining feature of human experience. Importantly this involves "an *affective* interaction between subject and object" and "the recognition of the significance of an object requires a relationship that is accompanied by a cognitive/ affective inner state" (Schore, 1994, p. 539). Such empirical studies have found that establishing a personally relevant universe of objects (Bollas, op. cit.) is essential for psychological and physiological health (Schore, 1994; Winnicott, 1971). Neuroscience also offers a physiological account of the importance of the management of the transition from dependency to individuality as proposed by Winnicott. In early life

the infant's interactions with its environment have been found to influence the evolution of brain structures which effect lifelong socio-emotional functioning. The communicative and emotional relationship between mother and child in early life is pivotal (Schore, 1994, p. 540).
4. Free associations can serve defensive functions as they may be part of the subject's efforts to avoid anxiety (Bollas, 2002, p.10).
5. In a pilot study I conducted I found that asking, "why do you like this film?" resulted in intellectualised answers. The interviewee would offer a critical evaluation of the text with responses such as "well-acted" or "good script" which told me nothing about the emotional significance of the film or the experience of viewing it. Once I substituted requests to "Tell me about..." I found a range of significant emotional and biographical narratives were generated.
6. The first interviews in this study were designed with only a small set of questions in mind including some biographical questions and a question about general likes and dislikes. Basic biographical questions concerning family of origin were incorporated at the start of the interview. I asked interviewees to tell me about family, where they grew up and went to school and their current life and employment. These questions obviously provide biographical context and also a sense of social trajectories, such as moving from a working-class family background into middle-class professional employment. In the second interview, I would ask the interviewee to "Tell me more about, and anything else that comes to mind about...?" This kind of open questioning was successful in generating biographical narratives, as the case studies show.

3 Spending Too Much Time Watching TV?

1. Office of National Statistics Labour Force Survey 2012 number of people living alone without spouse, partners or family rose from 6.6 million in 1996 to 7.6 million in 2012.
2. It is not uncommon periodically to find television associated with dystopian visions of modernity. See recent examples linking TV viewing with poor IQ, physical and emotional health: Parker-Pope http://www.nytimes.com/2008/03/04/health/04well.html (accessed 17 June 2012) and http://www.bbc.co.uk/news/health-14534025 (16 August 2012).
3. The American Heritage® Dictionary of the English Language, Fourth Edition copyright ©2000 by Houghton Mifflin Company. Updated in 2009. Published by Houghton Mifflin Company.
4. Collins English Dictionary – Complete and Unabridged© HarperCollins Publishers 1991, 1994, 1998, 2000, 2003.
5. This was done by special effects.

4 Favourite Things: Objects in the Life of a *Castaway*

1. For analysis of the Thatcher era see Hall and Jacques (1983) (eds.) *The Politics of Thatcherism* and Hall (1988) *The Hard Road to Renewal: Thatcherism and the Crisis of the Left.*

2. Ironically the collector is also a "bad" guy because he tries to keep the dolls in their boxes and even has one doll fully "mint in box" which *Woody* sees as working against the dolls' "true" nature of being used in play. Sophie of course has a large collection of "mint in box" dolls.
3. © Walt Disney Music Co. Ltd, Pixar Talking Pictures, Wide Music Inc., Walt Disney Music Company.
4. In the loft.
5. To triangulate my reading that Sophie's responses were not typical, I used the "Rate It" website, popular with people of Sophie's age, to assess the level of sadness of this scene. I was able to confirm that her response was not typical. In fact the majority of the responses found it rather corny or silly.
6. I have written permission to use this information which was double-checked, and a second written permission was given prior to publication of this book.
7. *Chuck* has a new *Wilson* ball sat next to him in his vehicle when he eventually gets back to his life and fulfils his promise to deliver the unopened parcel.

5 Mothers, Sons, Siblings and the Imaginative World of Working-Class Women's Viewing

1. A gothic horror starring Bette Davis. The story revolves around Charlotte, a middle-aged spinster who lives with guilt and fear after being fooled into thinking she murdered her fiancé in her youth. A friend and a cousin try to convince Charlotte she is mad in an attempt to force her into suicide so they can inherit her fortune.
2. Eastern European instructive fairy tale which concerns a conceited princess who is forced to confront her selfishness, which results in her love being turned into a bear. In the tale, to correct her mistakes she must undergo various trials (involving confrontation by numerous beasts). Eventually she learns compassion and kindness, ultimately saving her prince.
3. I have presumed that to a young child the perception was the "middle of the night", but this is likely to have been 10 or 11 o'clock.
4. Mary also explained that she was often terrified as she was convinced she could see snakes in her bedroom. Freudian theory would also link this with sexuality.
5. A film version of Thomas Hardy's *Jude*.
6. Mary does not even refer to her siblings by name.
7. Gerry Anderson puppet show. Marina is a mysterious mute woman who is like a mermaid, having human legs but she swims under the sea for long periods as a mermaid would. Her affinity is with the sea. She is also an object of desire for one of the male "agent" puppets.
8. It is possible to speculate on the phallic symbolism of trains and tunnels but I will not undertake this here.
9. The particular version Mary speaks about is a remake of a 1950 comedy. The film tells the story of George Banks, a regular middle-class American family man. His daughter graduates from college and announces her engagement. The humour revolves around his attempts to retain control of the finances and organisation of the wedding.

10. *Meet The Fockers* (2004, Dir. Jay Roach) is a comedy based on parents who disapprove of the family of the man their daughter is planning to marry.

6 Risky Viewing and Risky Method?: Psychoanalysis, Method and Defended Viewing

1. After a first interview Bill was taken into hospital, and so for logistical reasons we replaced the planned follow up interview with a series of letters. This has been shown to be an effective method in small scale research; see Denscombe (2007).
2. It is not clear what kind of student this woman was. I wondered if she was a student teacher, but none of the information I have clarifies this.
3. The Tavistock Model of Infant Observation draws on the ideas of John Bowlby and Esther Bick and is the quintessential learning from experience method. It involves a number of features which I adopted in the audience research I undertook, including recording inwardly not only what is seen but also the feelings that are generated. It is alert to the emotional tone of what is observed. Feelings and responses are written up along with other observations after the research encounter. Researchers are encouraged to "live with" the research experience and also discuss it with a supervisory team including affective responses (see Waddell, 2013).

References

Abraham, K. (1919) A Particular Form of Neurotic Resistance Against the Psycho-Analytic Method. In Bergman, R.S. and Hartman, F.R. (1990) (eds) (2nd ed.) *The Evolution of Psychoanalytic Technique*. Columbia: Columbia University Press. 82–88.

Abu-Lughood, L. (2002) Egyptian Melodrama – Technology of the Modern Subject? In Ginsburg, F.L., Abu-lughood, L. and Larkin, B. (eds) *Media Worlds: Anthropology on a New Terrain*. Berkeley, CA: University of California Press.

Andersen, S.M. and Berk, M.S. (1998) Transference in Everyday Experience: Implications of Experimental Research for Relevant Clinical Phenomena. In *Review of General Psychology*, Vol. 2, No. 1, 81–120.

Ang, I. (1985) *Watching 'Dallas': Soap Opera and the Melodramatic Imagination*. London: Methuen.

Ang, I. (1996) *Living Room Wars: Rethinking Media Audiences for a Postmodern World*. London: Routledge.

Ang, I. (2007) Television Fictions around the World: Melodrama and Irony in Global Perspective. In *Critical Studies in Television*, Vol. 2, No. 2,18–30.

Arato, A. and Gephardt, E. (1978) (eds) *The Essential Frankfurt School Reader*. London: Bloomsbury.

Bainbridge, C. and Yates, C. (2007) Everything to Play For: Masculinities, Trauma and the Pleasures of DVD Technology. In Bainbridge, C., Radstone, S., Rustin, M. and Yates, C. (eds) *Culture and the Unconscious*. London: Palgrave Macmillan.

Bainbridge, C., Ward, I. and Yates, C. (2014a) (eds) *Television and Psychoanalysis: Psycho-Cultural Perspectives*. London: Karnac.

Bainbridge, C., Ward, I. and Yates, C. (2014b) (eds) *Media and the Inner World: Psycho-Cultural Approaches to Emotion, Media and Popular Culture*. Basingstoke: Palgrave Macmillan.

Bartlett, F.C. (1932) *Remembering: A Study in Experimental and Social Psychology*. Cambridge: Cambridge University Press.

Bauman, Z. (1999) Modern Adventures of Procrastination. In *Parallax* Jan–March, Vol. 5, No. 1, Taylor and Francis.

Beck, U. (1992) *Risk Society*. London: Sage.

Beck, U. and Beck-Gernsheim, E. (2001) *Individualization*. London: Sage.

Beck, U., Giddens, A. and Lash, S. (1994) *Reflexive Modernization*. Cambridge: Polity Press.

Bellour, R. (1974/1975) *The Birds – Analysis of a Sequence*. London: BFI Education.

Bellour, R. (1988) Bergstrom Interview with Raymond Bellour. In Penley, C. (ed.) *Feminism and Film Theory*. London: BFI.

Bennett, T., Savage, M., Silva, E., Warde, A., Gayo-Cal, M. and Wright, D. (2009) *Culture, Class, Distinction*. London: Routledge.

Berg-Cross, G., Jennings, P. and Baruch, R. (1990) Cinematherapy: Theory and Application. In *Psychotherapy in Private Practice*, Vol. 8, No. 1.

Bettelheim, B. (1976) *The Uses of Enchantment: The Meaning and Importance of Fairy Tales*. New York: Random House.

Beynon, J. (2002) *Masculinities and Culture*. Buckingham: Open University Press.
Blumer, J.G. (1979) The Role of Theory in Uses and Gratifications Study. In *Communication Research*, Vol. 6, 9–36.
Blumer, J.G. & Katz, E. (eds) (1979) *The Uses of Mass Communications*. Beverly Hills: C.A. Sage.
Bion, W.R. (1962) (1984 ed.) *Learning From Experience*. London: Karnac.
Biressi, A. and Nunn, H. (2005) *Reality TV, Realism and Revelation*. London: Wallflower Press.
Bollas, C. (1987) *The Shadow of the Object*. London: Free Association.
Bollas, C. (1988) The Transformational Object. In Kohon, G. (ed.) *The British School of Psychoanalysis: The Independent Tradition*. London: Free Association. 83–101.
Bollas, C. (1992) *Being a Character: Psychoanalysis and Self Experience*. London: Routledge.
Bollas, C. (1995) *Cracking Up: The Work of Unconscious Experience*. London: Routledge.
Bollas, C. (2002) *Free Association*. Reading: Icon Books.
Bollas, C. (2009) *The Evocative Object World*. East Sussex: Routledge.
Bordwell, D. and Thompson, K. (1979) *Film Art: An Introduction*. United States: Addison Wesley.
Bourdieu, P. (1977) *Outline of a Theory of Practice*. Cambridge: Cambridge University Press.
Bourdieu, P. (1984) *Distinction: A Social Critique of the Judgment of Taste*. London: Routledge and Kegan Paul.
Bourdieu, P. and Wacquant, L.J.D. (1992) *An Invitation to Reflexive Sociology*. Chicago: University of Chicago Press.
Brunsdon, C. (2000) *The Feminist: The Housewife and The Soap Opera*. Oxford: Oxford University Press.
Butler, S. and Strupp, H. (1991) The Role of Affect in Time-Limited Dynamic Psychotherapy. In Safran, J. and Greenberg L. (eds) *Emotion, Psychotherapy and Change*. New York: Guildford Press.
Caldwell, J. (2003) Second Shift Media Aesthetics: Programming, Interactivity and User Flows. In Everett, A. and Caldwell, J. (eds) *New Media: Theories and Practices of Digitextuality*. London: Routledge.
Chaudhuri, A. (2002) Why Weepy Films Are Good For You. *Sunday Times London*. 21 April.
Chodorow, N. (1999) *The Power of Feelings*. New Haven: Yale University Press.
Clarke, S., Hahn, H. and Hoggett, P. (2008) (eds) *Object Relations and Social Relations: The Implication of the Relational Turn in Psychoanalysis*. London: Karnac. xii–xiii.
Coles, P. (2003) *The Importance of Sibling Relationships in Psychoanalysis*. London: Karnac.
Connell, R.W. (1995) *Masculinities*. Cambridge: Polity.
Corbin, J. and Strauss, A. (2008) (3rd ed.) *Basics of Qualitative Research*. Thousand Oaks, USA: Sage.
Cowie, E. (1997) *Representing the Woman: Cinema and Psychoanalysis*. Basingstoke: Macmillan.
Craib, I. (1998) *Experiencing Identity*. London: Sage.

Day Sclater, S. (2009) Introducing Psychosocial Studies of Emotion. In Day Sclater, S., Jones, D.W., Price, H. and Yates, C. (eds) *Emotion, New Psychosocial Perspectives*. Basingstoke: Palgrave Macmillan.

De Certeau, M. (1984) *The Practice of Everyday Life*. Berkeley: University of California Press.

Denscombe, M. (2007) *The Good Research Guide: Small Scale Social Research Project*. Berkshire: Open University Press.

Dreher, A. (2000) *Foundations for Conceptual Research in Psychoanalysis*. London: Karnac Books.

Ducey, A. (2009) More than a Job: Meaning, Affect and Training Health Care Workers. In Clough, P.T. and Halley, J. (eds) *The Affective Turn: Theorizing the Social*. Durham and London: Duke University Press.

Edwards, R., Hadfield, L., Lucey, L. and Mauthner, M. (2006) *Sibling Identity and Relationships: Sisters and Brothers, Relationships and Resources*. London: Routledge.

Elliott, P. (1974) Uses and Gratifications Research: A Critique and a Sociological Alternative. In Blumer J.G. and Katz, E. (eds) *The Uses of Mass Communications*. London: Sage. 249–268.

Elliott, A. and Lemert, C. (2006) *The New Individualism: The Emotional Costs of Globalization*. London: Routledge.

Elliott, A. and Urry, J. (2010) *Mobile Lives*. Abingdon Oxon: Routledge.

Ellis, J. (1989) *Visible Fictions*. London: Routledge.

Erickson, E. (1980) *Identity and the Life Cycle*. London and New York: W.W. Norton.

Erikson, E. (1959) (1994 ed.) *Identity and the Life Cycle*. London: W.W. Norton.

Evans, J. (2000) The Open University D853 *Identity in Question*, study Guide Block 2 'Psychoanalysis and Psycho-social Relations'. Milton Keynes: The Open University.

Faimberg, H. (2005) *The Telescoping of Generations*. Sussex: Routledge.

Featherstone, M. (2007) (2nd ed.) *Consumer Culture and Postmodernism*. London: Sage

Fiske, J. (1989a) *Reading the Popular*. USA: Unwin Hyman.

Fiske, J. (1989b) *Understanding Popular Culture*. USA: Unwin Hyman.

Fiske, J. (1992) The Cultural Economy of Fandom. In Lewis, L. (ed.) *The Adoring Audience*. London: Routledge.

Flood, M., Gardiner, J.K., Pease, B. and Pringle, K. (eds) *International Encyclopaedia of Men and Masculinities*. Abingdon: Routledge.

Foster, J. (2009) *Memory*. Oxford: Oxford University Press.

Freeman, M. (1989) Between the 'Science' and 'Art' of Interpretation: Freud's Methods of Interpreting Dreams. In *Psychoanalytic Psychology*, Vol. 6, No. 3, 293–308.

Freeman, M. (1993) *Re-Writing the Self: History, Memory, Narrative*. London: Routledge.

Freud, S. (1899) (1960 ed.) Screen Memories. In *The Standard Edition of the Complete Works of Sigmund Freud*, Vol. 3. New York: Basic Books. 47–69.

Freud, S. (1900 ed.) *The Interpretation of Dreams* Trans. By Strachey, J.S.E., Vol. iv and v, London: Hogarth Press, 1953, 1–627.

Freud, S. (1901) (2000 ed.) *The Psychopathology of Everyday Life*. London: Penguin.

Freud, S. (1913) (1985 ed.) *The Origins of Religion, Totem and Taboo and Other Works*. Translated by Strachey, J. and Dickson, A. (eds). London: Penguin.

Freud, S. (1914) Mourning and Melancholia. In Freud, S. Translated by Strachey, J. *Complete Works of Sigmund Freud,* Vol. XIV (1914–1916). London: Hogarth Press. 243–258.

Freud, S. (1915/2005) *The Unconscious*. Translated by Frankland. G. London: Penguin.

Freud, S. (1916) (1982 ed.) The Development of the Libido and Sexual Organisations. In Freud, S. Translated by Strachey, J. *Introductory Lectures on Psychoanalysis*. Reading: Penguin.

Freud, S. (1918) From the History of an Infantile Neurosis. In *SE*, Vol. 17, 1–122.

Freud, S. (1920) (1984 ed.) 'On Metapsychology: The Theory of Psychoanalysis', 'Beyond the Pleasure Principle', 'The Ego and the Id' and Other Works Translated by Strachey, J. and edited by Richards, A. Harmondsworth: Penguin.

Froggett, L. and Hollway, W. (2010) Psychosocial Research Analysis and Scenic Understanding. In *Psychoanalysis, Culture and Society*. Basingstoke: Palgrave Macmillan, Vol. 15, No. 3, 281–301.

Frosh, S. (2010) *Psychoanalysis Outside of the Clinic*. Basingstoke: Palgrave Macmillan.

Frosh, S. and Baraitser, L. (2008) Psychoanalysis and Psychosocial Studies. In *Psychoanalysis, Culture and Society*. Basingstoke: Palgrave Macmillan. December, Vol. 13, No. 4, 346–365.

Geraghty, C. (1991) *Women and Soap Opera*. Cambridge: Polity Press.

Gibbons, F. (1999) Films Unlock the Troubled Mind. *The Guardian London*. 3 August 1999.

Giddens, A. (1991) *Modernity and Self Identity: Self and Society in the Late Modern Age*. Cambridge: Polity Press.

Glaser, B.G. and Strauss, A.L. (1967) *The Discovery of Grounded Theory: Strategies for Qualitative Research*. New York: Aldine Gruyer.

Gomez, L. (1997) *An Introduction to Object Relations*. London: Free Association.

Gorton, K. (2009) *Media Audiences: Television, Meaning and Emotion*. Edinburgh: Edinburgh University Press.

Grant, J. and Crawley, J. (2002) *Transference and Projection*. Buckingham: Open University Press.

Gray, A. (1992) *Video Playtime: The Gendering of a Leisure Technology*. London: Routledge.

Grinberg, L. Sor, D. and Tabak de Bianchedi, E. (1993) *New Introduction to the Work of Bion*. New York: Jason Aronson.

Grolnick, S. (1990) *The Work and Play of Winnicott*. New Jersey: Jason Aronson.

Grossberg, L. (1992) Is There a Fan in The House? The Affective Sensibility of Fandom. In Lewis, L. (ed.) *Adoring Audience: Fan Culture and Popular Media*. London: Routledge.

Grosz, E. (1990) *Jacques Lacan: A Feminist Introduction*. London: Routledge.

Grotstein, J.S. (2002) 'Love Is Where It Finds You': The Caprices of the 'Aleatory Object'. In Scalia, J. (ed.) *The Vitality of Object*. Middletown USA: Wesleyan University Press.

Hadfield, L., Lucey, H., Mauthner, M. and Edwards, R. (2006) *Sibling Identity and Relationships*. London: Routledge.

Hall, S. (1980) Encoding/Decoding. In Hall, S., Hobson, D., Lowe, A. and Willis, P. (eds) *Culture, Media, Language: Working Papers in Cultural Studies, 1972–79*. London: Hutchinson. 128–138.

Hall, S. (1988) *The Hard Road to Renewal: Thatcherism and the Crisis of the Left*. London: Verso.

Hall, S. (1999) Introduction Looking and Subjectivity. In Evans, J. and Hall, S. (eds) *Visual Culture the Reader*. London: Sage, 309–314.

Hall, S. and Jaques, M. (1983) (eds) *The Politics of Thatchersim*. London: Lawrence and Wishart.

Hamilton, S. (2001) Film Star Treatment: Using the Power of Movies to Unlock the Mind. In *Daily Record*. 1 March.

Hanley, L. (2007) *Estates: An Intimate History*. London: Granta.

Happaney, K., Zelazo, P.D. and Stuss, D.T. (2004) Development of Orbitofrontal Function: Current Themes and Future Directions. In *Brain and Cognition*, Vol. 55, 1–10.

Harding, S. (1991) *Whose Science? Whose Knowledge?: Thinking from Women's Lives*. New York: Cornell University Press.

Harper, S. and Porter, V. (1996) Moved to Tears Weeping in the Cinema in Postwar Britain, *Screen*, Vol. 37, No. 2, 152–173.

Harrington, C.L and Bielby, D. (1995) *Soap Fans: Pursuing Pleasure and Making Meaning in Everyday Life*. Philadelphia: Temple University Press.

Harre, R. and Gillett, G. (1994) *The Discursive Mind*. London: Sage.

Henriques, J., Hollway, W., Urwin, C., Venn, C. and Walkerdine, V. (1984) *Changing the Subject*. London: Methuen.

Hesley, J.W. and Hesley J.G. (2000) *Rent Two Films and Let's Talk in The Morning*. New York: John Wiley and Sons.

Hills, M. (2002) *Fan Cultures*. London: Routledge.

Hills, M. (2005) Patterns of Surprise. The 'Aleatory Object' in Psychoanalytic Ethnography and Cyclical Fandom. In *American Behavioral Scientist*. March, Vol. 48, No. 7, 801–821.

Hinshelwood, R.D. (1989) *A Dictionary of Kleinian Thought*. London: Free Association.

Hoggett, P. (2008) What's in a Hyphen? Reconstructing Psychosocial Studies. In *Psychoanalysis, Culture and Society*, Vol. 13, No. 4, 379–384.

Hollway, W. (2009) Relationality: The Intersubjective Foundations of Identity. In Wetherell, M., Mohanty, C. and Mohanty, M. (eds) *Handbook of Identities*. London: Sage.

Hollway, W. (2010) Conflict in the Transition to Becoming a Mother: A Psycho-social Approach. In *Psychoanalysis, Culture and Society*, Vol. 15, No. 2.

Hollway, W. and Jefferson, T. (2000) (2nd ed.) *Doing Qualitative Research Differently: Free Association, Narrative and the Interview Method*. London: Sage.

Horton, D. and Whol, R. (1956) Mass Communication and Parasocial Interaction: Observations on Intimacy at a Distance. In *Psychiatry*, Vol. 19, No. 3, 188–211.

Hunt, J. (1989) *Psychoanalytic Aspects of Fieldwork. Qualitative Research Methods Series 18*. California Sage: USA.

Hurvich, M.S. (1989) Traumatic Moment, Basic Dangers and Signal Anxiety. In *Psychoanalytic Psychology*. USA: Lawrence Erlbauum Associates, Vol. 6, No. 3, 309–323.

Isaacs, S. (1952) The Nature and Function of Phantasy. In Klein, M., Heimann, P., Isaacs, S. and Riviere, J. (eds) *Developments in Psycho-Analysis*. London: Hogarth. 67–122.

Jefferson, T. (2008) "What Is the Psychosocial"? A Response to Frosh and Baraitser. In *Psychoanalysis, Culture and Society*, Vol. 13, No. 4, 366–373.

Jemstedt, A. (2011) Introduction. In Bollas, C. (ed.) *The Christopher Bollas Reader*. London: Routledge. xii–xxxviii.

Jenson, J. (1992) Fandom as Pathology: The Consequences of Characterization. In Lewis, L. (ed.) *Adoring Audience: Fan Culture and Popular Media*. London: Routledge.

Johnson, R., Chambers, C., Raghuram, P. and Tincknell, E. (2004) *The Practice of Cultural Studies*. London: Sage.

Jordan, T. (2008) Security in the Social: Gardens and Harry Potter. In Carter, S., Jordan, T. and Watson, S. (eds) *Security: Sociology and Social Worlds*. Manchester: Manchester University Press. 17–46.

Katz, E., Blumer, J.G. and Gurevitch, M. (1974) Uses and Gratifications Research. In *The Public Opinion Quarterly*, Vol. 37, No. 4, 509–523.

Kavka, M. (2008) *Reality Television, Affect and Intimacy*. Basingstoke: Palgrave Macmillan.

Kernberg, O. (1976) *Object Relations Theory and Clinical Psychoanalysis*. New York: Jason Aronson.

Klein, M. (1929) (1986 ed.) Infantile Anxiety Situations Reflected in a Work of Art and in the Creative Impulse. In Mitchell, J. (ed.) *The Selected Melanie Klein*. London: Hogarth Press. 84–95.

Klein, M. (1940) Mourning and Its Relation to Manic Depressive States. In Frankiel, R. (ed.) *Essential Papers on Object Loss*. New York: New York University Press. 95–122.

Klein, M. (1946) Some Notes on Some Schizoid Mechanisms. In Mitchell, J. (1986) (ed.) *The Selected Melanie Klein*. New York: Free Press.

Klein, M. (1948) A Contribution to the Theory of Anxiety and Guilt. In *International Journal of Psycho-analysis*, Vol. 29, 14–23.

Klein, M. (1960) *The Psychoanalysis of Children* Trans. Alix Strachey New York: Grove press Originally published by Hogarth Press and The Institute of Psychoanalysis no 11, International Psycho-analytical Library London.

Klinger, B. (2006) *Beyond the Multiplex, Cinema, New Technologies and the Home*. London and Berkeley: University of California Press.

Kuhn, A. (1995) *Family Secrets: Acts of Memory and Imagination*. London: Verso.

Kuhn, A. (2013) *Little Madnesses: Winnicott, Transitional Phenomenon and Cultural Experience*. London: I.B. Tauris.

Lacan, J. (2004) *The Four Fundamental Concepts of Psycho-Analysis*. London: Karnac.

Lacan, J. (2007) *Ecrits*. New York: W.W. Norton.

Lacey, J. (1997) *Seeing Through Happiness: Class, Gender and Popular Film: Liverpool Women Remember the 50's Film Musical*. Unpublished PhD thesis Goldsmiths College, University of London.

Lacey, J. (1999) Hollywood Musicals and the construction of the American Dream in Liverpool in the 1950's. In Kuhn, A. and Street, S. (eds) *Journal of Popular British Cinema* Kuhn, Vol. 2. Great Britain: Flicks Books.

Langan, R. (2014) *The American Journal of Psychoanalysis*, 74, 21–30. Doi: 10.1057/ajp.2013.36 available at http://www.palgrave-journals.com/ajp/journal/v74/n1/full/ajp201336a.html [accessed 4 March 2014]

Laplanche, J. (1992) Notes on Afterwardsness, Translated by Stanton, M. In Fletcher, J. and Stanton, M. (eds) *Jean Laplanche: Seduction, Translation, Drives*. London: ICA. 217–223.

Laplanche, J. (1999) Notes on Afterwardsness. In Laplanche, J. *Essays on Otherness*. Fletcher, J. (ed.) London: Routledge.

Laplanche, J. and Pontalis, J.B. (1973) Fantasy and the Origins of Sexuality In *International Journal of Psychoanalysis*, Vol. 49, Part 1. 1–19.

Laplanche, J. and Pontalis, J.B. (1973) *The Language of Psychoanalysis*. London: Karnac.

Lapsley, R. and Westlake, M. (1988) *Film Theory*. Manchester: Manchester University Press.

Lasch, C. (1979) *The Culture of Narcissism*. New York: W.W. Norton.

Layton, L. (2002a) Cultural Hierarchies, Splitting and the Heterosexist Unconscious. In Fairfield, S., Layton, L. and Stack, C. (eds) *Bringing the Plague: Towards a Postmodern Psychoanalysis*. New York: Other Press.

Layton, L. (2002b) Psychoanalysis and the 'Free Individual'. In *Journal of Psycho-Social Studies*, Vol. 1, No. 1 [E-Journal] http://www.btinternet.com/-psycho-social/vol1/JPSS_LL1.htm.

Layton, L. (2003) That Place Gives Me the Heebie Jeebies. In *Critical Psychology*, No. 10, 36–49.

Layton, L. (2004) A Fork in the Road: On 'Defining' the Unconscious and Its Stakes for Social Theory. In *Psychoanalysis, Culture and Society*, Vol. 9, 33–51.

Lopate, C. (1977) Daytime Television: You'll Never Want to Leave Home. In *Radical America*, Vol. 11, No. 1, 33–51.

Lorenzer, A. (1986) Tiefenhermeneutische Kulturanalyse. In Lorenzer, A. (ed.) *Kultur*. Fischer: Taschenbuch Verlag.

Mann, G. (2002) Transformational, Conservative and Terminal Objects: The Application of Bollas's Concepts to Practice. In Scalia, J. (ed.) *The Vitality of Objects*. Middletown, USA: Wesleyan University Press.

Mansfield, N. (2000) *Subjectivity*. New York: New York University Press.

Matthews, M. (1998) The Body as Instrument. In *Journal of British Association of Psychotherapists*, Vol. 35, 17–36.

Mayne, J. (1993) *Cinema and Spectatorship*. London: Routledge.

McGuigan, J. (1992) *Cultural Populism*. London: Routledge.

McLeod, J. (1997) *Narrative and Psychotherapy*. London: Sage.

McQuail, D. (1987) *Mass Communication Theory*. London: Sage.

Melanie Klein Trust (2014) *Internal Objects*. Available at http://www.melanie-klein-trust.org.uk/internal-objects (accessed March 2014).

Miller, D. (2008) *The Comfort of Things*. Cambridge: Polity Press.

Mishler, E.G. (1986) *Research Interviewing: Context and Narrative*. Cambridge, MA: Harvard University Press.

Mitchell, J. (1986) *The Selected Melanie Klein*. New York: Free Press.

Mitchell, J. (2003) *Siblings, Sex and Violence*. Cambridge: Polity/Blackwell.

Modleski, T. (1982) *Loving With a Vengeance: Mass Produced Fantasies for Women*. Hamden, CT: Shoestring Press.

Morley, D. (1992) *Television Audiences and Cultural Studies*. London: Routledge.
Mulvey, L. (1975/1988) Visual Pleasure and Narrative Cinema. In Penley, C. (ed.) (1988) *Feminism and Film Theory*. London: BFI.
Mumford, L.S. (1995) *Love and Ideology in the Afternoon: Soap Opera, Women and Television Genre*. New York: John Wiley and Sons.
Neale, S. (1986) Melodrama and Tears. In *Screen*. Nov–Dec., Vol. 27, No. 6.
Nelson-Jones, R. (2001) *The Theory and Practice of Counseling and Therapy*. London: Continuum.
Office of National Statistics. *Labour Market Survey 2012*. Available at
http://www.ons.gov.uk/ons/rel/lms/labour-market-statistics/march-2012/index.html (accessed 08 August 2014).
Ogden, T. (1989) (2004 ed.) *The Primitive Edge of Reason*. New Jersey: Jason Aronson and Karnac.
Ogden, T. (1990) *The Primitive Role of Experience*. Northvale, NJ: Jason Aronson.
Ogden, T. (2004) On Holding and Containing, Being and Dreaming. In *International Journal of Psychoanalysis*, Vol. 85, 1349–1364.
Ogden, T. (1999) *Reverie and Interpretation: Sensing Something Human*. London: Karnac.
Olesen, V. (2000) Feminisms and Qualitative Research At and Into the Millennium. In Denzin, N. and Lincoln, Y. (eds) *Handbook of Qualitative Research*. Thousand Oaks, USA: Sage.
Ouellette, L. and Hay, J. (2008) *Better Living Through Reality TV*. Oxford: Blackwell.
O'Shaughnessy, M. (1994) Promoting Emotions: Understanding Films, Understanding Ourselves. In *Metro Media and Education Magazine*, Vol. 97.
O'Shaughnessy, M. (1999) (1st ed.) *Media and Society*. Australia: Oxford University Press.
Pace, S. (2012) Writing the Self into Research: Using Grounded Theory Analytic Strategies in Autoethnography. In *Text*: *Special Issue: Creativity: Cognitive, Social and Cultural Perspectives* April 2012. Available at http://www.textjournal.com.au/speciss/issue13/Pace.pdf (accessed 5 March 2014).
Penley, C. (ed.) (1988) *Feminism and Film Theory*. London: BFI.
Peske, N. and West, B. (2002) *Advanced Cinematherapy. The Girl's Guide to Finding Happiness One Movie at a Time*. New York: Dell Press.
Phillips, A. (2002) Futures. In Scalia, J. (ed.) *The Vitality of Objects*. Middletown, USA: Wesleyan University Press.
Phillips, A. (2007) *Winnicott*. London: Penguin.
Polkinghorne, D.E. (1995) Narrative Configuration in Qualitative Analysis. In Hatch, J.A. and Wisniewski, R. (eds) *Life History and Narrative*. London: Falmer Press.
Polkinghorne, D.E. (1998) *Narrative Knowing and the Human Sciences*. New York State: University of New York Press.
Postman, N (1985) *Amusing Ourselves to Death*. London: Methuen.
Radstone, S. (ed.) (2000) Introduction and Screening Trauma: Forest Gump, Film and Memory. In Radstone, S. (ed.) *Memory and Methodology*. Oxford: Berg.
Redman, P. (2009) Affect Revisited: Transference-countertransference and the Unconscious Dimensions of Affective, Felt and Emotional Experience. In *Subjectivity*. Palgrave: Macmillan, Vol. 26, 51–68.
Redman, P. and Whitehouse-Hart, J. (2008) 'I Just Wanted Her Out' Attachment, the Psycho-social and Media Texts. In Redman, P. (ed.) *Attachment, Sociology and Social Worlds*. Manchester: Open University Press.

Reiss, D. (1981) *The Family's Construction of Reality*. Cambridge, MA: Harvard University Press.

Ricoeur, P. (1988) *Time and Narrative*. Vol. 3. Chicago: University of Chicago Press.

Roberts, B. (2002) *Biographical Research*. Milton Keynes: Open University Press.

Rose, G. (1989) *Visual Methodologies*. London: Sage.

Rose. N. (1999) (2nd ed.) *Governing the Soul: The Shaping of the Private Self*. London: Free Association.

Rosenbaum, M. (1998) Childhood 'Screen Memories' Are They Forgotten? In *Psychosomatics*. Jan–Feb., Vol. 39, No. 1, 68–71.

Rustin, M. (2009) 'The Missing Dimension: Emotions in the Social Sciences. In Day Sclater, S., Jones, D.W., Price, H. and Yates, C. (eds) *Emotion New Psychosocial Perspectives*. Basingstoke: Palgrave Macmillan.

Rustin, M. and Rustin, M. (2001) *Narratives of Love and Loss in Modern Children's Fiction*. London: Karnac.

Sandler, P.C. (2005) *The Language of Bion*. London: Karnac.

Savin-Williams, R.C. (2001) Memories of Same Sex Attachment. In Kimmell, M.S. and Messner, M.A. (eds) *Men's Lives*. USA: Allyn and Bacon.

Scannel, P. (2007) What Happens When I Turn on the Television Set? In Nyre, L. *Westminster Papers in Communication and Culture*. London: University of Westminster, Vol. 4, No. 2, 24–35.

Schafer, R. (1981) Narration in the Psychoanalytic Dialogue. In Mitchell, W.T.J. (ed.) *On Narrative*. Chicago: University of Chicago Press.

Schore, A. (2003) *Affect Regulation and the Repair of the Self*. New York: W.W. Norton.

Schore, A. (2010) The Right Brain Implicit Self: A Central Mechanism of the Psychotherapy Change Process. In Petrucelli, J. (ed.) *Knowing, Not Knowing and Sort of Knowing: Psychoanalysis and the Experience of Uncertainty*. London: Karnac. 177–201.

Schwandt, T. (2000) Three Epistemological Stances for Qualitative Inquiry: Interpretivism, Hermeneutics and Social Constructionist. In Denzin, N. and Lincoln, Y. (eds) *Handbook of Qualitative Research*. Thousand Oaks, USA: Sage.

Scott, A. (1996) *Real Events Revisited: Fantasy Memory and Psychoanalysis*. London: Virago.

Segal, J. (1992) *Melanie Klein*. London: Sage.

Seiter, E. (1990) Making Distinctions: Case Study of a Troubling Interview. In *Cultural Studies*, Vol. 4, No. 1, 61–84.

Sennett, R. and Cobb, J. (1972) *The Hidden Injuries of Class*. Cambridge: Cambridge University Press.

Sharpe, S.A. and Rosenblatt, A.D. (1994) Oedipal Sibling Triangles. In *Journal of American Psychoanalytic Association*, Vol. 42, 499–573.

Silverstone, R. (1994) *Television and Everyday Life*. London: Routledge.

Skeggs, B. (1997) *Formations of Class and Gender*. Thousand Oaks, London: Sage.

Skeggs, B. (2005) *Formations of Class and Gender: Becoming Respectable*. London: Sage.

Smith, D. (2000) The Mirror Image of the Present. In *Psychoanalytische Perspectieven*, NR. 39. Available at http://www.psychoanalyticscheperspectieven.be/wp-content/uploads/2012/10/screen-memories.pdf (accessed 20 December 2013).

Smith, M. (1995) *Engaging Characters; Fiction, Emotion and the Cinema*. Oxford: Oxford University Press.

Spence, D. (1982) *Narrative Truth and Historical Truth*. New York: Norton.
Spezzano, C. (1993) *Affect in Psychoanalysis*. Hillsdale, NJ: The Analytic Press.
Stacey, J. (1988) Desperately Seeking Difference. In Gamman, L. and Marshment, M. (eds) *The Female Gaze*. London: The Women's Press. 112–129.
Stacey, J. (1994) *Star Gazing: Hollywood Cinema and Female Spectatorship*. London: Routledge.
Suler, J. (2009) E-mail Transference: Seeing the Other Clearly. In *On Line Magazine* http://www.selfhelpmagazine.com./articles/internet/emailtransference.html (accessed 26 November 2009).
Sutton, P. (2004) Afterwardsness in Film. In *Journal for Cultural Research*, Vol. 8, No. 3, 385–405.
Symington, N. (1988) *The Analytic Experience: Lectures from the Tavistock*. London: Free Association Press.
Tester, K. (1998) Bored and Blasé: Television, the Emotions and Georg Simmel. In Bendelow, G. and Williams, S. (eds) *Emotions in Social Life: Critical Themes and Contemporary Issues*. London: Routledge. 83–96.
Tolson, A.V. (2011) I'm Common and My Talking Is Quite Abrupt' (Jade Goody): Language and Class in Celebrity Big Brother. In Skeggs, B. and Wood, H. (eds) *Reality Television and Class*, London: BFI Publications.
Tolson, A.V. (2006) *Media Talk: Spoken Discourse on TV and Radio*. Edinburgh: Edinburgh University Press.
Thompson, J.B. (1995) *The Media and Modernity: A Social Theory of the Media*. Cambridge: Polity Press.
Thompson, K. (1988) *Breaking the Glass Armour: Neo-formalist Film Analysis*. Princeton, NJ: Princeton University Press.
Volkan, V.D. (1972) The Linking Objects of Pathological Mourners. In *Archives of General Psychiatry*. August, Vol. 27.
Volkan, V.D. and Ast, G. (2014) *Siblings in the Unconscious and Psychopathology*. London, Karnac.
Waddell, M. (2013) Infant Observation in Britain: A Tavistock Approach Infant observation. *International Journal of Infant Observation and Its Applications*, Vol. 16, No. 1 4–22.
Walkerdine, V. (1986) Video Replay. Families, Films and Fantasy. In Burgin, V. Donald, J. and Kaplan, C. (ed.) *Formations of Fantasy*. London: Routledge.
Walkerdine, V. (1993) Daddy's Gonna Buy You a Dream to Cling to (and Mummy's Gonna Love You Just as Much as She Can): Young Girls and Popular Television. In Buckingham, P. (ed.) *Reading Audiences: Young People and the Media*. Manchester: Manchester University Press.
Walkerdine, V. (1997) *Daddy's Girl: Young Girls and Popular Culture*. Basingstoke: Macmillan.
Walkerdine, V. (2008) Contextualizing Debates about Psychosocial Studies. In *Psychoanalysis, Culture and Society*, Vol. 13, No. 4, 41–345.
Walkerdine, V. and Blackman, L. (2001) *Mass Hysteria, Critical Psychology and Media Studies*. Basingstoke: Palgrave.
Walkerdine, V., Melody, J. and Lucey, H. (2001) *Growing up Girl: Psychosocial Explorations of Class and Gender*. Basingstoke: Sage.
Warner, M. (1995) *From the Beast to the Blonde: On Fairytales and Their Tellers*. London: Chatto and Windus.
Wasko, J. (2001) *Understanding Disney*. Cambridge: Polity Press.

Wengraf, T. (2001) *Qualitative Research Interviewing: Biographic Narrative and Semi-Structured Methods*. Los Angeles and London: Sage.
Whitehead, S.M. (2002) *Men and Masculinities Key Themes and New Directions*. Cambridge: Polity Press.
Whitehouse-Hart (2014) Programmes for People Who are Paranoid about the Way they Look': Some Thoughts on Paranoia, Mirrors and Contemporary Lifestyle Television. In Bainbridge, C., Ward, I. and Yates, C. (eds) *Emotion and Culture*. Basingstoke: Palgrave Macmillan.
Williams, R. (1974) *Television, Technology and Cultural Form*. London: Fontana.
Willis, P. (1982) *Learning to Labour: How Working Class Kids Get Working Class Jobs*. Morningside Edition. New York: Columbia University Press.
Winnicott, D.W. (1951) 'Transitional Objects and Transitional Phenomena' in Winnicott, D.W *Through Paediatrics to Psychoanalysis Collected Papers* London: Karnac, 229–242.
Winnicott, D.W. (1965a/1990) Ego Distortion in Terms of True and False Self. In Winnicott, D.W. (ed.) *The Maturational Processes and the Facilitating Environment*. chapter 12. London: Karnac. 140–153.
Winnicott, D.W. (1965b) The Development of the Capacity for Concern. In *The Maturational Processes and the Facilitating Environment*. London: Karnac. 73–83.
Winnicott, D.W. (1965c) The Theory of the Parent-Infant Relationship. In *The Maturational Processes and the Facilitating Environment*. London: Karnac. 37–55.
Winnicott, D.W. (1971) *Playing and Reality*. London: Routledge.
Winnicott, D.W. (1990) *The Maturational Processes and the Facilitating Environment*. London: Karnac.
Wood, H (2009) *Talking With Television: Talk Shows and Modern Self-Reflexivity*. Illinois: University of Illinois Press.
Wood, H., Skeggs, B. and Thumin, N. (2008) Oh Goodness I am Watching Reality TV: How Methodology Makes Class in Multi-method Audience Research. In *European Journal of Cultural Studies*, Vol. 11, No. 1, 5–24.
Woods, J. (1993) Repeatable Pleasures: Notes on Young People's Use of Video. In Buckingham, D. (ed.) *Reading Audiences: Young People and the Media*. Manchester: Manchester University Press.
Yin, R.K. (1989) *Case Study Research: Design and Methods*. Applied Social Research Methods Series. Vol. 5. Newbury Park, London: Sage.
Yin, R.K. (2012) *Applications of Case Study Research*. Thousand Oaks, USA: Sage.
Young, R. (2009) *Potential Space: Transitional Phenomena*. Online Paper http://human-nature.com/rmyoung/papers/paper55.html (accessed 6 October 2009).
Zittoun, T. (2006) *Transitions Development Through Symbolic Resources*. Greenwich, CT: Information Age Publishing.
Zizek, S. (1989) *The Sublime Object of Ideology*. London: Verso.

Index

Printed and bound in the United States of America